SpringerBriefs in Environmental Science

SpringerBriefs in Environmental Science present concise summaries of cutting-edge research and practical applications across a wide spectrum of environmental fields, with fast turnaround time to publication. Featuring compact volumes of 50 to 125 pages, the series covers a range of content from professional to academic. Monographs of new material are considered for the SpringerBriefs in Environmental Science series.

Typical topics might include: a timely report of state-of-the-art analytical techniques, a bridge between new research results, as published in journal articles and a contextual literature review, a snapshot of a hot or emerging topic, an in-depth case study or technical example, a presentation of core concepts that students must understand in order to make independent contributions, best practices or protocols to be followed, a series of short case studies/debates highlighting a specific angle.

SpringerBriefs in Environmental Science allow authors to present their ideas and readers to absorb them with minimal time investment. Both solicited and unsolicited manuscripts are considered for publication.

More information about this series at http://www.springer.com/series/8868

Dania Abdul Malak · Katriona McGlade
Diana Pascual · Eduard Pla

Adapting to Climate Change

An Assessment of Vulnerability and Risks
to Human Security in the Western
Mediterranean Basin

 Springer

Dania Abdul Malak
European Topic Centre-University of
   Malaga
Malaga
Spain

Katriona McGlade
Ecologic Institute
Berlin
Germany

Diana Pascual
CREAF
Barcelona
Spain

Eduard Pla
CREAF
Barcelona
Spain

ISSN 2191-5547   ISSN 2191-5555   (electronic)
SpringerBriefs in Environmental Science
ISBN 978-3-319-51678-3   ISBN 978-3-319-51680-6   (eBook)
DOI 10.1007/978-3-319-51680-6

Library of Congress Control Number: 2016961334

Printed on acid-free paper

This Springer imprint is published by Springer Nature
The registered company is Springer International Publishing AG
The registered company address is: Gewerbestrasse 11, 6330 Cham, Switzerland

Preface

The Mediterranean Basin is one of the world's regions most vulnerable to climate change. It is already experiencing the increasingly severe effects of a changing climate such as higher average temperatures and the incidence of extreme events such as unprecedented heat waves, severe droughts, and major floods. These impacts threaten environment, society, and economy and can place human security at risk.

This book examines the water-related impacts of climate and global change in the protected area of the UNESCO Intercontinental Biosphere Reserve of the Mediterranean (hereafter IBRM). The IBRM straddles the continents of Africa and Europe and spans two countries (Spain and Morocco). The Reserve presents a particularly interesting case for study: on the one hand, predicted climate change effects are rather homogenous across both sides of the IBRM; on the other hand, the level of vulnerability and consequences for human security vary considerably depending on the national jurisdiction. This is due to the differing socio-economic contexts and patterns of land-use in Spain and Morocco, as well as the capacity of their respective policy frameworks to mitigate environmental threats.

This book considers the case of the IBRM through the application of a new approach for assessing vulnerability. Three distinct assessments (hydro-ecological, socio-economic, and policy) are conducted on the basis of a range of inputs (a hydro-ecological model, water management and climate change scenarios, desk-based research, and consultation with expert stakeholders). These individual assessments are then considered as a whole to produce an integrated vulnerability assessment for the IBRM. Key research insights are accompanied by recommendations for reducing vulnerability and ensuring human security within the IBRM and the region more generally.

The findings in this book signal that when climate change is coupled with other compounding factors—such as a paucity of structural water resources, unsustainable socio-economic growth, or inadequate policy frameworks—it can lead to serious consequences for human security (and its components e.g., water, food, personal, and economic security). The analysis suggests that the IBRM would benefit from improved implementation of current national-level climate change and water management policies, with due consideration to the interplay between different sectoral strategies e.g., agriculture and tourism. Above all, it is estimated that the Reserve could better develop practical management approaches adapted to local needs that integrate the local population and their know-how e.g., on the use of drought-resistant crops to find climate adaptation solutions that reduce vulnerability and risks to human security.

Malaga, Spain   Dania Abdul Malak
Berlin, Germany   Katriona McGlade
Barcelona, Spain   Diana Pascual
Barcelona, Spain   Eduard Pla

Acknowledgements

The authors would like to take the opportunity to thank all the persons who have contributed to this study. The work would not have been possible without the involvement and valuable contributions of stakeholders from Spain and Morocco, who took the time to share their views and constructively discuss the effects of climate change on the IBRM. Their inputs provided us with a depth and range of insights that were fundamental to our understanding of socio-ecological systems and vulnerabilities in the region.

We would also like to show our gratitude for the involvement and cooperation with the Government of Morocco (Haut Commissariat aux Eaux et Forêts) and the Regional Government of Andalusia. We thank both for their support, their contributions to discussions at the focus groups and expert meetings, and for the provision of data and maps.

We would like to thank Steffen Schwörer, whose research on water policy in Spain provided the initial foundation for this aspect of Chap. 4 and to Ana Iglesias for her review. We also thank Ms. Hanane Nauoi (Parc National de Talassemtane) for her valuable support in developing the focus groups and the provision of data, and J.A. Lopez-Bustins (Physic Geography and AGR Department, University of Barcelona) for helping with climatic analysis and climatic data. We are thankful to Mr. Eric Mino from the Euro-Mediterranean Information System on know-how in the Water sector (EMWIS) for his support and willingness to provide us with updates to the existing literature. We would also like to show our gratitude to Mr. Jaume Fons for his contribution to the creation of adapted soil maps for use with the hydrological model.

Finally, we thank our colleagues and families for their support during the writing process, as well as the Directorate General for Research and Innovation of the European Commission, whose 7th Framework Programme funded the Climate Change, Hydro-conflicts, and Human Security (CLICO) project and the primary research in this book.

Contents

List of Figures

List of Tables

List of Boxes

Definitions

Adaptation: The process of adjustment to actual or expected climate and its effects. In human systems, adaptation seeks to moderate or avoid harm or exploit beneficial opportunities. In some natural systems, human intervention may facilitate adjustment to expected climate and its effects (IPCC 2014).

   Adaptive capacity: The ability of systems, institutions, humans, and other organisms to adjust to potential damage, to take advantage of opportunities, or to respond to consequences (IPCC 2014).

   Climate change: Climate change refers to a change in the state of the climate that can be identified (e.g., by using statistical tests) by changes in the mean and/or the variability of its properties and that persists for an extended period, typically decades or longer. Climate change may be due to natural internal processes or external forcings such as modulations of the solar cycles, volcanic eruptions, and persistent anthropogenic changes in the composition of the atmosphere or in land use. Note that the Framework Convention on Climate Change (UNFCCC), in its Article 1, defines climate change as: "a change of climate which is attributed directly or indirectly to human activity that alters the composition of the global atmosphere and which is in addition to natural climate variability observed over comparable time periods." The UNFCCC thus makes a distinction between climate change attributable to human activities altering the atmospheric composition, and climate variability attributable to natural causes (IPCC 2014).

   Exposure: The presence of people, livelihoods, species or ecosystems, environmental services and resources, infrastructure, or economic, social, or cultural assets in places that could be adversely affected (IPCC 2014).

   Flood: The overflowing of the normal confines of a stream or other body of water, or the accumulation of water over areas that are not normally submerged. Floods include river (fluvial) floods, flash floods, urban floods, pluvial floods, sewer floods, coastal floods, and glacial lake outburst floods (IPCC 2012).

   Global change: Global change refers to planetary-scale changes in the Earth system. More completely, the term "global change" encompasses: planetary-scale changes to atmospheric circulation, ocean circulation, climate, the carbon cycle, the nitrogen cycle, the water cycle and other cycles, sea-ice changes, sea-level changes, food webs, biological diversity, pollution, health, fish stocks, and more. Civilization is now a large driver of global change, so the term includes population, the economy, resource use, energy, development, transport, communication, land use and land cover, urbanization, and globalization (IGBP 2015).

Hazard: The potential occurrence of a natural or human-induced physical event that may cause loss of life, injury, or other health impacts, as well as damage and loss to property, infrastructure, livelihoods, service provision, and environmental resources (IPCC 2012).

Human security: A condition that exists when the vital core of human lives is protected, and when people have the freedom and capacity to live with dignity. This includes the universal and culturally specific, material and non-material elements necessary for people to act on behalf of their interests. Many phenomena influence human security, notably the operation of markets, the state, and civil society (IPCC 2014).

Impacts: The effects of extreme weather and climate events and of climate and global change on natural and human systems e.g., loss of life, livelihoods, health, ecosystems, economies, societies, cultures, services, and infrastructure (Adapted from IPCC 2014).

Mitigation (of climate change): A human intervention to reduce the sources or enhance the sinks of greenhouse gases (GHGs). It also includes to reduce the sources of other substances which may contribute directly or indirectly to limiting climate change, including, for example, the reduction of particulate matter emissions that can directly alter the radiation balance (e.g., black carbon) or measures that control emissions of carbon monoxide, nitrogen oxides, volatile organic compounds, and other pollutants that can alter the concentration of tropospheric ozone which has an indirect effect on the climate (IPCC 2014).

Risk: The potential for consequences where something of value is at stake and where the outcome is uncertain, recognizing the diversity of values. Risk is often represented as probability or likelihood of occurrence of hazardous events or trends multiplied by the impacts if these events or trends occur. In this report, the term risk is often used to refer to the potential, when the outcome is uncertain, for adverse consequences on lives, livelihoods, health, ecosystems and species, economic, social and cultural assets, services (including environmental services), and infrastructure (IPCC 2014).

Scenario: A scenario is a coherent, internally consistent, and plausible description of a possible future state of the world. It is not a forecast; rather, each scenario is one alternative image of how the future can unfold. A projection may serve as the raw material for a scenario, but scenarios often require additional information (e.g., about baseline conditions). A set of scenarios is often adopted to reflect, to the extent possible, the range of uncertainty in projections. Other terms that have been used as synonyms for scenario are "characterisation," "storyline," and "construction" (Pinkerton and Rom 2014).

Sensitivity: The degree to which a system or species is affected, either adversely or beneficially, by climate variability or change. The effect may be direct (e.g., a change in crop yield in response to a change in the mean, range, or variability of temperature) or indirect (e.g., damages caused by an increase in the frequency of coastal flooding due to sea-level rise) (IPCC 2007a).

Tipping point: A critical threshold at which the future state of a system can be qualitatively altered by a small change in forcing (Lenton et al. 2008). Applied to climate change, it is defined as a threshold for abrupt and irreversible change, remain uncertain, and that the risk associated with crossing multiple tipping points increases with rising temperature (IPCC 2014).

Vulnerability: The propensity or predisposition to be adversely affected. Vulnerability encompasses a variety of concepts including sensitivity or susceptibility to harm and lack of capacity to cope and adapt. A broad set of factors such as wealth, social status, and gender determine vulnerability and exposure to climate-related risk (IPCC 2014).

Chapter 1
Introduction

Abstract Climate change—a strand of global change—is having increasingly severe impacts on natural and human systems in the Mediterranean region. These impacts include both incremental changes such as higher average temperatures as well as extreme events like unprecedented heat waves, severe droughts, and major flood events. Regional climate scenarios show an intensification of these trends, which—coupled with growing freshwater demand—are expected to negatively impact rural livelihoods and increase urban migration in the Mediterranean. However, the extent to which socio-ecological systems are vulnerable depends on their capacity to respond and adapt. Policy frameworks are a key way to reduce vulnerabilities, support the adaptive capacity of natural and human systems and to ensure human security. This can be done for example by: improving the socio-economic situation of particular population groups and thus their capacity to adapt to changing conditions; improving the state of the natural environment, thus limiting the impact of climate change; or by protecting populations against specific risks (e.g. floods) that endanger human security. The development and effectiveness of these policy frameworks varies across the region and depends on a range of socio-cultural, historical and economic factors.

Keywords Mediterranean basin · Spain · Morocco · Climate change · Climate adaptation · Adaptive capacity · Human security · Food security · Water security · Agriculture · Migration · Floods · Water scarcity · Drought · Policy frameworks · Vulnerability

1.1 Existing Impacts of Global and Climate Change on the Mediterranean Region

Global change refers to planetary-scale changes in the Earth system that affect, *inter alia*, the globe's climate, atmospheric and ocean circulation, the carbon, nitrogen, water and nitrogen cycles, sea-ice, sea-levels, food webs, fish stocks and biological diversity. The changes to these systems are simultaneous, unprecedented in their

© The Author(s) 2017   1
D. Abdul Malak et al., *Adapting to Climate Change*,
SpringerBriefs in Environmental Science, DOI 10.1007/978-3-319-51680-6_1

rate and magnitude, and are increasingly driven by human activities. Global change affects the provision of basic goods and services such as food, water, clean air and an environment conducive to human health, with serious implications for vulnerable populations, ecosystems and associated services (The Amsterdam Declaration on Global Change 2001; IPCC 2001; World Bank 2014; IGBP 2015). In the Mediterranean region, global change dynamics include population growth, agricultural intensification, economic and social development, tourism and over consumption of already scarce water resources (ENPI 2007; Benoît and Comeau 2005). These dynamics are increasing pressures on water resources especially in coastal zones (Fornés et al. 2005) and on environmental flows, which are essential to the function of aquatic ecosystems (EEA 2012). In this book, we focus primarily on climate change—a strand of global change—referring to the altered composition of the global atmosphere, beyond natural variability, which takes place as a consequence of human activity (IPCC 2014). In particular, we focus on vulnerabilities to water-related impacts of climate change. The Mediterranean region has been subject to major climate change impacts over recent decades as a result of decreased precipitation and increased temperature (EEA 2012). From 1961 to 1990 a warming of 0.2 °C per decade has been observed in the region. Since this period the Mediterranean has been warming at an even faster rate with an increasing rise in monthly summer temperature (World Bank 2012). An analysis of twentieth century long-term changes in the water cycle of Mediterranean land areas also highlights a weak, albeit significant, long-term negative precipitation trend over Mediterranean land areas (Mariotti et al. 2008).

The Mediterranean region is known for the diversity of its sub-humid, arid and semi-arid ecosystems, which are the result of fluctuations in temperature, wind, moisture availability and soil composition over very short distances (Zamora et al. 2007). These ecosystems provide services with tangible economic value to local populations such as crop and livestock production, tourism, water, watershed protection and energy supply. These systems are naturally adaptable and do have a certain capacity to adjust to stressful conditions, including slight changes in climate (Tielbörger et al. 2014). However, the pressures of global change are stretching this capacity to its limit and increasing species extinction rates, desertification, and rapid collapse of ecosystem functionality at local and regional level (IPCC 2002).

Drought is a global "creeping phenomenon" that occurs in high as well as in low rainfall areas (Gillette 1950). Different types of drought exist, but all types are caused by a deficiency in precipitation leading to water shortage for activities such as plant growth, or for socio-economic sectors such as agriculture or tourism (Wilhite and Glantz 1985). Globally, drought phenomena can have serious economic, social, and environmental consequences such as land degradation, human migration, famine, disease and loss of human life (IPCC 2012). The Mediterranean region has already suffered significant increases in drought episodes since 1970 (Iglesias and Moneo 2005; Agoumi 2003; Vogt and Somma 2000; Wilhite and Vanyarkho 2000) causing huge economic damages over the last two decades (CRED 2010, 2014). This increasing frequency further adds to the existing challenges of sustainable water resource management in a region already affected by natural water scarcity (Iglesias et al. 2007; Hisdal et al. 2001). There is also growing evidence of limited capacity to cope with socio-economic and agricultural demands in periods of drought (Scheffran and Battaglini 2011).

The lack of water in rural areas is leading to short term—and occasionally long term—migratory flows towards urban centres where water supply is usually prioritised and non-agricultural employment can be found (ENPI 2007). In parallel, regional water demand and consumption in coastal zones, is estimated to have doubled in the second half of the twentieth century (Plan Bleu 2008). Non-renewable water resource use in the Mediterranean e.g. from fossil aquifers is close to its limit and potentially renewable resources are being overexploited beyond their capacity, leading to periodic water shortages (Schnepf and Lutter 2012). The overuse of resources puts food, water, energy, and environmental security at risk (COM 2016). In some cases this is leading to conflicts over water resource distribution and rights among individual users as well as countries (Touchan et al. 2010; CRED 2010, 2014; Agoumi 2003; Parish and Funnell 1999).

Floods are defined as the overflowing of the normal confines of a stream or other body of water, or the accumulation of water over areas that are not normally submerged. Flood types include river (fluvial) floods, flash floods, urban floods, pluvial floods, sewer floods and coastal floods (IPCC 2012). The impacts of flooding depend on a range of factors such as settlement patterns and land use (IPCC 2007a). Direct impacts can be severe and include injury, loss of life and damage to infrastructure. Secondary impacts include the contamination of water supply and increased risk of water borne diseases (IPCC 2012). There is already evidence of changing trends in regional flood events in the Mediterranean with a higher frequency of less intensive rainy days coupled with an increase in intense precipitation episodes along the Mediterranean coast (Alpert et al. 2002). Population increases and changes in land use such as the trend of building on floodplains have increased the exposure of coastal populations and have resulted in significant personal economic losses as well as loss of life (Perez et al. 2016; Barrera-Escoda and Llasat 2015).

1.2 Future Impacts of Global and Climate Change in the Mediterranean Region

The extent to which climate change affects individual regions will vary over time and according to the ability of different environmental systems to mitigate or adapt to change (IPCC 2007a). Climate projections indicate that the Mediterranean region— especially North Africa and the Middle East—is a global "hotspot" for worsening extreme heat, flood, drought, and aridity conditions as an effect of climate change (Giorgi 2006; Navarra and Tubiana 2013). Projections for the Mediterranean region indicate that more than 90% of summers will have highly unusual heat extremes in a scenario of a 4 °C global temperature increase, compared to between 20 and 40% of summers in a world with 2 °C warming (World Bank 2014a). It is very probable that average annual temperature increases will be slightly higher than the global average (Plan Bleu 2008), being more pronounced in North Africa (Navarra and Tubiana 2013). An increase in temperature would lead to profound changes at the level of natural habitats leading to losses in biodiversity and ecosystem service provisions (Klausmeyer and Shaw 2009).

In the Mediterranean region, and more specifically in the Middle East and North Africa, the effects of 2 and 4 °C warming estimated from regional models also foresee a substantial reduction in precipitation (Fig. 1.1). Under the scenario of a 2 °C temperature increase, the annual river discharge levels are projected to drop by more than 15% (IPCC 2007b). Reduced precipitation and consequently water availability may cause declining yields, seriously affect food production in the Mediterranean, for irrigation and rainfed farming in the Mediterranean (FAO 2011). Together with an expected reduction in precipitation, drought episodes will also increase (Kerr 2005). This may compromise future freshwater availability in the Mediterranean region, an area that is already subjected to water stresses and recurrent water shortages (Magnan et al. 2009). Growing water demands due to population increase and the growth of economic sectors are likely to exacerbate existing water scarcity in the region (Garrido and Iglesias 2007); as well as its associate health risks (Murray et al. 2012).

Parallel to an overall decrease in precipitation, the Mediterranean region will also see a greater level of variability and intensity of rainfall. Additional negative impacts resulting from flooding are projected by the end of the 21st century both inland and in low altitude coastal zones. Increased storm intensity and flash

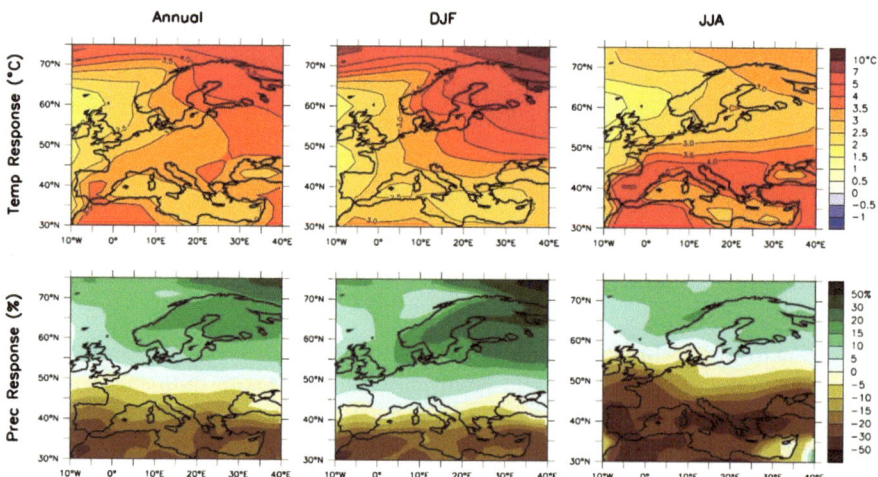

Fig. 1.1 Temperature and precipitation changes over the Mediterranean from the MMD-A1B simulations. First row: Multi-model annual DJF and JJA temperature change between 1980 to 1999 and 2080 to 2099, averaged over 21 models. Second row: same as top, but for fractional change in precipitation (IPCC 2007b)

flooding are expected to accelerate runoff and soil erosion affecting beaches, infrastructure and coastal sectors such as tourism (IPCC 2007a) as well as posing further threats to agriculture and risks to human and environmental systems (Agoumi 2003). Unfortunately, there is as yet only low confidence in numerical projections of flood magnitude or frequency resulting from climate change and there is controversy regarding regional models and data sets used for projections (Pryor et al. 2009), especially in the Mediterranean region (Turco and Llasat 2011; Kundzewicz et al. 2013).

Agriculture is a major source of employment for many parts of the southern Mediterranean meaning that livelihoods are heavily dependent on natural and semi natural environments (FAO 2011). Future climate change is expected to increase the risk to livelihoods and subsistence farming systems, causing dramatic direct effects to farmers' income, and threatening the availability of food for local communities. Furthermore, crop yield declines coupled with impacts in other grain-producing regions could contribute to increasing food prices in the Mediterranean region, where a growing food import dependency further exacerbates such risks (World Bank 2014a). In coastal areas, high exposure to sea-level rise in the coming decades is linked to groundwater salinisation of soils and reduction in land productivity, creating 'uninhabitable' areas (Zghibi et al. 2013).

Climate change impacts negatively affect rural livelihoods dependent on natural systems and can contribute to large-scale internal and international migration and rural land abandonment in favour of urban centres (ENPI 2007). However, it is difficult to decouple this type of migration from other forms of voluntary migration such as labour migration. Estimates on the number of affected people over the coming decades differ considerably, depending on the extent of existing measures to adapt to climate and environmental change, future demographic and economic trends, and related levels of affluence and poverty (IOM 2009; Parry et al. 2007; Stern 2006; UNU-EHS 2007). Nevertheless, around 20 million people are estimated to have been displaced globally in 2008 as a consequence of extreme weather events (IOM 2009) and increasing pressures on the environment and impacts of environmental hazards are likely to become major root causes for increased migration in the future. As Europe is the highest net receiver of international migration globally (UNU-EHS 2007) the continent will likely be impacted by any such increase.

1.3 Adaptive Capacity, Vulnerability and Human Security

Desertification, water scarcity, drought and flooding all have multidimensional implications not only for natural systems, but also for human populations and local economies in the Mediterranean region (Kostopoulou et al. 2014). However, the extent to which socio-ecological systems are vulnerable to these impacts is defined by the capacity and preparedness of these systems to respond and adapt (IPCC 2007a; Karas 1998). In this way, both the impacts of climate change and the adaptive capacity are key determiners for the level of vulnerability and potential consequences for human security. It is the assumption of this book that the primary responsibility for reducing vulnerabilities and ensuring human security lies with governments and thus a strong focus is placed on the role that policy frameworks play in supporting adaptive capacity. This can be done for example by: improving the socio-economic situation of particular population groups and thus their capacity to adapt to changing conditions; improving the state of the natural environment, thus limiting the impact of climate change; or by protecting populations against specific risks (e.g. floods) that endanger human security.

In the Mediterranean, policy frameworks support adaptive capacity, reduce vulnerability and ensure human security in different ways and to differing degrees, reflecting the socio-cultural, historical and economic diversity of the region. Mediterranean countries that are members of the European Union have adopted *acquis communautaire* that—for the most part—ensure a shared basic level of policy development, implementation and monitoring in terms of protection of the environment, populations and their human rights. For example Spain has a comprehensive and sustainable approach to drought and flood risk management in place that in some cases has already shown its capacity to reduce vulnerabilities. EU Member States are also industrialised countries with higher levels of GDP that allows for a greater dedication of resources to the implementation of these policies. In non-EU Mediterranean countries, the picture is varied, but in developing contexts, there may be fewer resources to ensure effective implementation.

This book focuses on Morocco and Spain and does not attempt to draw conclusions on the Mediterranean region as a whole with regards to policy development and implementation. There are clear differences between these two countries, which despite being subjected to similar physical impacts have markedly different policy frameworks and adaptive capacities. Whatever the approach taken, climate projections highlight the urgent need for all countries in the region to develop policies and programmes supporting adaptive capacity at all levels of governance to reduce the effects of global change to human security and its components such as water, food, personal and economic security.

This book begins by introducing the physical and socio-economic impacts of climate change on the Intercontinental Biosphere Reserve of the Mediterranean (IBRM) (Chap. 2). The following two chapters contain the research base for the analysis: in Chap. 3 a hydro-ecological assessment and a socio-economic assessment are carried out to produce an impact assessment for the IBRM; while in Chap. 4, an assessment of policy frameworks is conducted. These three assessments are then brought together in an integrated assessment of vulnerability (Chap. 5). Chapter 6 integrates the conclusions of the assessments and recommends potential action points to reduce vulnerability and ensure human security in the IBRM.

Chapter 2
The Intercontinental Biosphere Reserve of the Mediterranean

Abstract The creation of the Intercontinental Biosphere Reserve of the Mediterranean (IBRM) is a joint endeavour between Spain and Morocco; countries with very similar natural and biophysical conditions, but different socio-economic contexts and policy frameworks. Water related impacts of climate change in the region mean that on the one hand, increasing levels of drought are heightening inter-sectoral competition for water resources, and on the other hand, more intense and recurrent floods are leading to soil erosion, biodiversity loss, and negative consequences for ecosystem services. There is a marked socio-economic contrast between Spain and Morocco with the Moroccan side of the IBRM being particularly dependent on natural resources and exposed to climate change impacts. However, water-related impacts have the potential to affect human security on both sides of the IBRM. This was evidenced, for example, by the extensive droughts of the 1990s which had substantial effects on the agricultural economy in both Spain and Morocco, threatening food and water security and provoking migratory flows to urban centres. In this way, unfavourable effects of global change within the IBRM also affect society, economy and human security in the wider region.

Keywords Mediterranean basin · Spain · Morocco · Climate change · Floods · Water scarcity · Drought · Human security · Food security · Water security · Agriculture · Migration

2.1 Background

The Intercontinental Biosphere Reserve of the Mediterranean (hereafter IBRM) is located on the Western side of the Mediterranean Sea basin within the coordinates 4°41′24″ W, 36°57′29″ N (NW) and 5°50′25″ W, 34°47′38″ N (SE) (Fig. 2.1). The Reserve straddles the continents of Europe and Africa, passing through the Strait of Gibraltar and covering a surface of around one million hectares. The Reserve was created in 2006, in a joint endeavour between Spain and Morocco under the UNESCO Man and Biosphere Reserve Programme. This international cooperation

© The Author(s) 2017

D. Abdul Malak et al., *Adapting to Climate Change*,

SpringerBriefs in Environmental Science, DOI 10.1007/978-3-319-51680-6_2

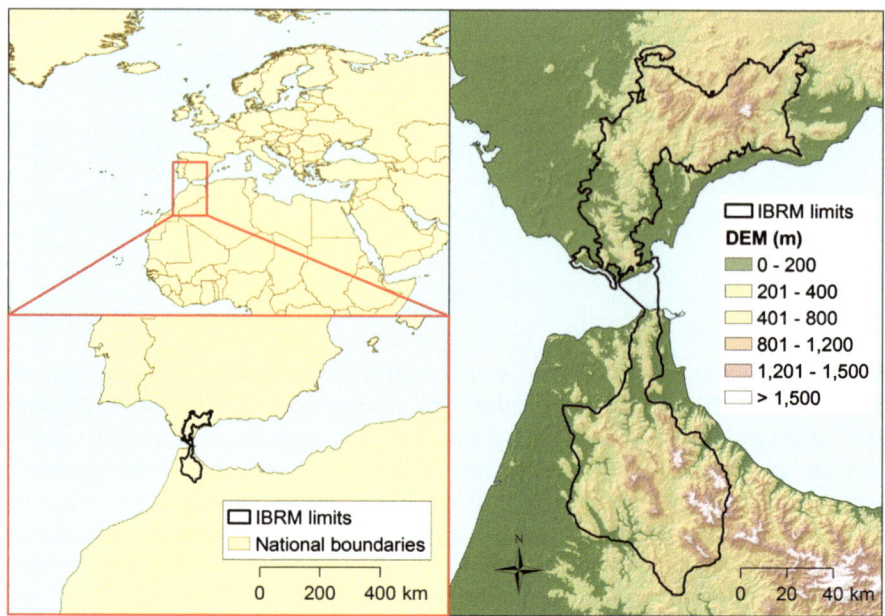

Fig. 2.1 Location of the IBRM (*left*). Detail of the IBRM boundaries (*right*), digital elevation model (DEM in meters)

involves the creation of a common management model for the sustainable territorial development of the Reserve, the preservation of the shared ecosystem and the conservation of the region's natural diversity and cultural heritage (Molina and Villa 2008). A Memorandum of Understanding on the Coordination of the IBRM was signed between the Spanish and Moroccan governments in 2011 and an Action Plan (2011–2015) was developed for the Reserve.

The IBRM is located in the semi-arid zone of the Mediterranean Basin where the climatic conditions are known for their seasonal contrasts, characterised by a high annual variability in precipitation (500–800 mm/year) and frequent episodes of drought. Due to its geographic location, the IBRM is influenced by both the winds of the western Mediterranean Sea and cold fronts from the Atlantic. These wet fronts are responsible for the discharge of a high percentage of the Reserve's annual precipitation, increasing the annual precipitation to around 2000 mm/year in some parts of the Reserve such as Grazalema (Spain). In addition, the high mean annual temperature (16–19 °C) and the high solar radiation in the region lead to extensive evapotranspiration, increasing the threat of drought, especially during the summer months.

Although both sides of the IBRM share very similar natural and biophysical conditions, the impact of human activities has shaped the landscape very differently resulting in diverse land use patterns: 70% of the Spanish side is covered by forest areas, scrubland and grassland while on the Moroccan side this amounts to around

29% (GlobCover v.2.3. 2009, 2010). In parallel, cropland covers around 39% of the Spanish side but accounts for more than 60% of coverage on the Moroccan side, while water bodies and artificial areas cover around 1% of the area in the Spanish side of the reserve, and 0.25% of the Moroccan side of the Reserve. Differing land use patterns are also reflected in the extension of protected areas: 70% of the Spanish side of the Reserve is protected in contrast to 30% of the Moroccan side (Molina and Villa 2008).

The Reserve has 556,359 inhabitants within its boundaries (INE 2016) but the reference population (including surrounding towns that share a border contact with the IBRM) is 719,293 inhabitants. Rapid socio-economic growth and technological transformation in the areas surrounding the IBRM have taken place in recent decades. For example, on the Spanish border of the IBRM, the Costa del Sol, has seen a high level of urban expansion, while near the Moroccan border, the opening of the Tanger-Med cargo and passenger port in 2007 has led to a substantial increase in human pressures. High demand for freshwater services for tourism and irrigation can lead to excessive water abstraction and the overexploitation of water resources (Llamas and Martinez-Santos 2005; Iglesias et al. 2007) as well as increasing land degradation. The demand for freshwater, energy and food is also expected to increase significantly over the next decades under the pressure of population growth, economic development, trade, urbanisation, and climate change (Hoff 2011). At the same time, regional projections indicate that precipitation will become scarcer (Magnan et al. 2009). If socio-economic development is carried out in an unsustainable manner, and given the paucity of structural water resources in the IBRM region, there is a potential that the wider hydrological cycle could be significantly affected.

2.2 Impacts of Climate Change on the Physical Environment

In the Rif region on the Moroccan side of the IBRM, researchers concur that the latter half of the twentieth century was one of the driest in the last nine centuries, causing net declines in dam water reserves, deficits in groundwater resources and limitations to drinking and irrigation water supply (Parish and Funnell 1999; Agoumi 2003; Touchan et al. 2010). After the 1970s, the decreasing trend in precipitation levels in Morocco provoked water deficiency in many areas (Ouassou et al. 2005). In the last decades, the already frequent incidence of droughts in the IBRM region has shown a significant tendency to increase (Karaky and Arndt 2002; INE 2016). These drought events have negative impacts on water quantity and quality. They affect the production of hydroelectric power due to declines in dam water levels, lead to interruptions in drinking water treatment plant services and affect the healthy functioning of a range of ecosystems, resulting in fish deaths, water that is unsuitable for human consumption and the increase of waterborne diseases (Bennis and Sadeq 1998; Ameziane et al. 2001).

Bruggeman et al. (2010) developed climate change scenarios for the IBRM that point to a maximum rainfall decrease of 24–25% in 2040–2069, compared to 1961–1990, with summer and autumn periods more affected (A1B scenario). The same scenario projects a temperature increase of 2.4–2.5 °C for the same period. Even under the lower emission B1 scenario, the projected temperature increase is expected to be between 1.7 and 1.8 °C. These climatic stresses are expected to deeply affect water resource availability by decreasing water runoff and aquifer recharge. Spain's Mediterranean basins are already under very high stress and will have to meet a growing part of demand from "unconventional" sources by 2025, such as water re-use and desalination processes (Blinda et al. 2006). Andalusia currently reuses around 21 hm^3 water per year and desalinates 23 hm^3 per year accounting for 1.9% and the 2.1% respectively of total available water resources within the region (AAA 2011). During the same period, at a national level, Morocco will still be within the limits of its renewable water resources; however, the country may experience localised or exceptional stress in some areas given the unequal distribution of water resources (Blinda et al. 2006).

At the other end of the spectrum, floods—and flash floods in particular—have significant impacts on the Reserve. Greater storm intensity is expected to accelerate runoff, increasing soil erosion (Agoumi 2003), which is further exacerbated by deforestation and inappropriate agricultural practices. Annual losses of soil exceeding 4000 Mg/km^2 have been registered in the region of the Rif in Northern Morocco (Ouassou et al. 2005) and are threatening 21% of the soil of the province of Malaga in Southern Spain (RME 2008). As a consequence, a massive loss of biodiversity is expected, which—linked with projected biodiversity loss as a direct consequence of human pressures—could imply a drastic reduction in associated ecosystem services in the near future (Garcia et al. 2010; Magnan et al. 2009). Examples of impacts to ecosystem services from soil erosion include decreased freshwater supplies, soil productivity, forest productivity, plant pollination and fish stocks. Furthermore, although flash floods have significant impacts within the area of the Reserve, their effects also increase downstream, impacting not only the IBRM, but the coastal regions outside its boundaries.

2.3 Impacts of Climate Change on Socio-economic Sectors

While the impacts of climate change can have significant consequences for society and economy, human pressures and global change are also responsible for compounding these climatic effects. Sectors such as agriculture and tourism are threatening freshwater ecosystems such as the Guadalhorce river basin on the Spanish side of the IBRM, which is at risk of not meeting the environmental objectives of the Water Framework Directive (Carrasco et al. 2008). These human pressures negatively impact freshwater ecosystem services, such as fisheries, water purification and flood prevention and what is more, have a knock-on effect for the poorest local communities in the IBRM (Garcia et al. 2010). Climate change is expected to intensify these pressures in the Mediterranean basin (Magnan et al. 2009) threatening water quality and quantity and the sustainability of socio-economic activities in the IBRM.

Within the IBRM region, the vulnerability of the regional population is highly linked to the climate and its fluctuations since the main economic activities (agriculture and tourism) are strongly dependent on water availability (Agoumi 2003). However, there is a marked socio-economic contrast between Spain and Morocco as depicted in Table 2.1 which shows a higher level of dependency on natural resources in Morocco than in Spain, and thus a higher vulnerability to climate change effects.

In Andalusia, on the Spanish side of the IBRM, people involved in primary sectors e.g. agriculture, are also identified as the main vulnerable groups, followed by those working in tourism (Mestre-Barceló 1995; Roberts 2002). Although the agricultural sector is smaller on the Spanish than on the Moroccan side, agriculture is seen to have an important role in maintaining economic activity and employment in rural areas (Eurobarometer 2014). Recurrent droughts have led to serious losses and injuries to flora and fauna in forestry, cropping and pastoral sectors on the Spanish side of the IBRM. In particular cereals, vines, and olive trees are among the

Table 2.1 Some socio-economic indicators for Morocco and Spain

Indicators	Morocco	Spain
GDP/per capita in US Dollar (2009 est.)[a]	4600	33,700
Per capita—by occupation:[b] • Agriculture • Industry • Services	18.8% 32.6% 48.6% (2009 est.)	3.4% 26.9% 69.6% (2009 est.)
Labour force—by occupation:[c] • Agriculture • Industry • Services	44.6% 19.8% 35.5% (2006 est.)	4% 26.4% 69.5% (2008 est.)
Population living in rural areas (%)	47	22.4 (in Andalusia)
Population living in Rural area in IBRM (%)	93	–
Land ownership (average farm side in ha.)	2–3	35
Use of total water resources (%)[d]	43	32
Water resources per capita (m^3/person/year)[e]	895	2557
Water use for agriculture (%)	86	70
Water use for tourism (%)	–	13
Irrigated agriculture (%)	10	90

[a]http://www.theodora.com/wfbcurrent
[b]http://www.theodora.com/wfbcurrent
[c]http://www.theodora.com/wfbcurrent
[d]Millennium Development Goal indicators 2000
[e]http://earthtrends.wri.org

most affected agricultural crops. The tourism sector is affected by shortages in water supply to numerous villages, adding to the effect of higher extremes in summer temperatures (Méndez 2008).

In Spain, agricultural production is primarily commercial; whereas in Morocco, and indeed within that side of the IBRM, traditional agriculture and extensive farming act as the main means of subsistence for a large proportion of the population (Table 2.1). Unfavourable weather conditions are the main sources of risk to subsistence farming systems, especially in marginal land, where drought years can affect directly farmers' income. The Moroccan population has adopted two main strategies to avoid negative impacts to their livelihoods and human security in recent decades: (1) the development of unsustainable agricultural practices (overgrazing, illegal cannabis cropping in higher gradient slopes, forest overexploitation) with a high environmental impact in terms of soil degradation (Moore et al. 1998; Barrow and Hicham 2000); and (2) the migration from rural areas to coastal cities (Bennis and Sadeq 1998). Periodic drought episodes and significant rainfall variability do not only affect subsistence farming, but also agricultural production and its contribution to gross domestic product (GDP) more generally (Ameziane et al. 2001; Karaky and Arndt 2002).

Water-related impacts on the physical environment and on socio-economic sectors have the potential to affect human security on both sides of the IBRM. For example, the extensive droughts in Spain, Morocco and Tunisia in the mid-1990s impacted food production in rain-fed agriculture (i.e. primarily Morocco) as well as irrigation-dependent systems (i.e. primarily Spain). During this time, cereal production dropped dramatically from 9.5 million tons in 1994 to 1.6 million tons in 1995 (Skees et al. 2001) and many systems had cease production due to insufficient water reserves (CRED 2010). In Spain, the effects of the 1990s droughts on the agricultural economy were severe, with over 6 million people affected, many of whom moved from rural areas to cities to ensure drinking water supply (personal communication, stakeholders' meeting 2012).

The IBRM is one of the key migratory routes to Europe. The Moroccan side of IBRM has received migrating populations from other parts of Africa for many years, placing additional pressure on the region's society, economy and infrastructure. In addition, during especially dry periods, inhabitants of the IBRM, living in rural areas and with greater vulnerability to climate change, add to the movement towards urban centres and in some cases towards the EU (ENPI 2007). This migratory trend has been particularly intense during the region's frequent periods of poor harvests (Refass 1992). Migration is further expected to increase as a possible consequence of climate change, (Fermin 2009a, b) although exact estimates are hindered by the complex interplay of economic and political conditions, social and cultural context. The possible rise in sea water level implies groundwater salinising the soils may lead to even further reasons for land abandonment and the creation of "uninhabitable" places (Zghibi et al. 2013). These migratory patterns have important effects on the rural economy of the IBRM which would be negatively affected by any permanent loss of local knowledge and human resources.

Chapter 3
Impact Assessment

Abstract At present, most global change impact assessments have been under-taken for distinct sectors. The Intergovernmental Panel on Climate Change has however stressed the need for integrated approaches to provide an overview of the cumulative effects of climate change. This impact assessment is multi-disciplinary in its approach and combines two distinct assessments to analyse the impacts of global change on the Intercontinental Biosphere Reserve of the Mediterranean that straddles Morocco and Spain. In a first step, regional climate change scenarios are defined and used to carry out a hydro-ecological assessment based on modelling the physical impacts of climate change. In a second step, water management scenarios are defined and used to conduct a socio-economic assessment based on empirical evidence from stakeholders on the effects of scenarios on regional livelihoods and adaptive measures taken. The results of these assessments are then analysed and overall impacts of climate change on physical and socio-economic systems are identified for the Intercontinental Biosphere Reserve of the Mediterranean.

Keywords Mediterranean basin · Spain · Morocco · Climate change · Climate adaptation · Adaptive capacity · Human security · Food security · Water security · Floods · Water scarcity · Drought · Tourism · Agriculture · Hydro-ecological model · Scenarios · Impact assessment

3.1 Approach and Methodology

The impact assessment conducted in this book is multi-disciplinary in its approach. In a first step, a set of regional climate change scenarios are defined. These scenarios are then applied in two separate assessments: a hydro-ecological assessment and a socio-economic assessment. The hydro-ecological assessment is based on modelling the physical impacts of climate change. The socio-economic assessment is based on climate change scenarios and empirical evidence from stakeholders

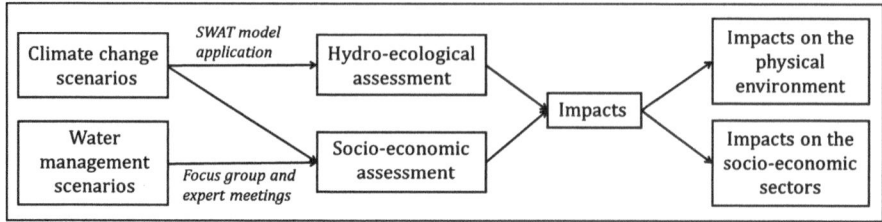

Fig. 3.1 Conceptual overview of the impact assessment framework applied in the IBRM

regarding the effects of water management scenarios on regional livelihoods and adaptive measures taken. The outputs of these assessments are considered together and on this basis the overall impacts of climate change on the IBRM are identified and analysed. The overall process is summarized in the schematic Fig. 3.1.

3.2 Climate Change Scenarios

The impacts of anthropogenic emissions in the IBRM are assessed with climate projections. For this, input data consisted of an ensemble median of different global circulation models (GCMs) under A2 IPCC scenario as provided in Bruggeman et al. (2010). The IPCC has described four scenario families (A1, A2, B1, and B2). These depend on the induced effects of carbon emissions associated with development, defined by the demographic, economic and technologic development (IPCC 2007a). The A2 scenario predicts high levels of anthropogenic emissions, accompanied by high economic and regional development (Parry et al. 2007). These projections consist of the climate change signals for average annual temperature and precipitation over the period 2040–2069, contrasted with the reference period of 1961–1990. The A2 scenario results describe an average decrease in precipitation of 17–18%, and an increase in annual temperature of 2.3–2.4 °C.

Using this climate change signal, a future climate series was built. This was accomplished using the weather generator developed for the GOTILWA+ model system (Keenan et al. 2010). The GOTILWA+ model is capable of generating future daily climate series (1961–2070), incorporating the signal of the different climate change scenarios and also taking into account statistical patterns of the observed climate. The GOTILWA+ system handles cumulative uncertainties associated with the GCMs' projections, downscaling procedures, land-use assumptions, or models used to be taken into account in the analysis and discussion of results (Ewen et al. 2006; Beven 2011).

3.3 Hydro-Ecological Assessment

Simulation of hydro-ecological responses to climate change was conducted using the Soil and Water Assessment Tool (SWAT) (Arnold et al. 1998). This model is semi-distributed and physically-based, capable of estimating surface and subsurface flow, erosion, sediment deposition, and nutrient movements at the basin scale at a daily time step temporal resolution (Gassman et al. 2007). SWAT has been a popular tool for a variety of applications related to water resources, but it has rarely been applied to the study of climate change impacts in Mediterranean watersheds (Nunes et al. 2008).

For an appropriate simulation of actual hydrological processes, the basin was divided into the sub-basins through which streams are routed. Sub-units within sub-basins are organized based on the combination of slope, soil, and land use characteristics considered as hydrologically homogeneous; these are referred to as hydrologic response units (HRUs). Importantly, calculations within the model are performed at the level of the HRU. The study of the IRBM only utilized basins that were wholly or partly within the Reserve.

For simulation of the hydrological regime, the SWAT model required input data on climate, topography, soil, and land use. Climate series, based on data from 23 meteorological stations in watersheds within and from the surrounding area of the IRBM (8 in Andalusia and 15 in Morocco, provided by the Spanish State Meteorological Agency, AEMET, and the Loukkos Hydrological Basin Agency), included precipitation and minimum and maximum temperatures. These series were filled and corrected for the effects of altitude on precipitation and temperature by SWAT equations. Elevation data was obtained from different sources for each country: Andalusian elevation data was obtained from the SRTM digital elevation model (90 m spatial resolution, NASA), and Moroccan data was obtained from the ASTER global digital elevation model (ASTER GDEM, 30 m, Earth Remote Sensing Data Analysis Centre). Soil data, compiled specifically for the project, was based on the Andalusian Soil Map (1:400,000, Andalusian Agricultural Regional Ministry and CSIC) and the Moroccan Geomorphological map (1:100,000, Mohamed V University in Rabat). Land use data in Andalusia was obtained from the Andalusia Use and Land Cover Map (2007, 1:25,000, Andalusian Environmental Regional Ministry), and land use data in Morocco was obtained from Globcover V (2005–2006 version, 300 m, ESA GlobCover Team) (Table 3.1).

The SWAT model parameters were then adjusted and calibrated to the specific characteristics of the Reserve, using historical data. Next, the prediction capacity of the model was measured by comparing simulated results and measured data (from a different time period). Using historical stream flow and reservoir management data from the periods 1983–2009 (Andalusia) and 1981–2007 (Morocco), the SWAT model was calibrated and validated at a monthly time step. For this procedure, data was available from 16 Andalusian gauging stations, 3 Moroccan gauging stations, and 6 Andalusian reservoirs. Due to the partial availability and low quality of

Table 3.1 Data used in SWAT model, resolution or scale of the data and data source

Data	Country	Resolution/scale	Data source
Elevation (m)	Spain	90 m	SRTM digital elevation model NASA
	Morocco	30 m	ASTER global digital elevation model (ASTER GDEM) Earth Remote Sensing Data Analysis Centre
Soil type	Spain	1:400,000	Andalusian Soil Map (Andalusian Agricultural Regional Ministry and CSIC)
	Morocco	1:100,000	Moroccan Geomorphological map Mohamed V University in Rabat
Land use	Spain	1:25,000	Andalusia Use and Land Cover Map (2007) Andalusian Environmental Regional Ministry
	Morocco	300 m	Globcover V (2005–2006 version) ESA GlobCover Team

stream flow and reservoir data series, model calibration and validation was done for 3-year periods. There were three main target objectives during calibration and validation: (1) simulated curves should be similar to measured ones; (2) mean flow values and total contributions should be similar between simulated and measured data; (3) good numerical statistics for the Nash—Sutcliffe efficiency coefficient (NSE) and the Root Mean Square Error (RMSE)—observations standard deviation ratio (RSR), as indicated in Moriasi et al. (2007). The equations used in calculations of statistics and performance ratings are shown in Table 3.2.

Figure 3.2 and Table 3.3 show calibration and validation outputs for daily stream flow for two gauging stations in the IBRM. Similar results to these were obtained for other gauging stations modelled in the context of this assessment. Beginning with the graphical comparison of simulated and measured results (Fig. 3.2), the model shows good fit as well as agreement in mean daily values. Secondly, NSE and RSR statistics (Table 3.3) adjusted in a satisfactory manner in all gauging stations with the exception of Khrofa gauging station (Morocco).

Table 3.2 Equations for estimation of NSE and RSR statistics, where Y_i^{obs} is the ith observation for the constituent being evaluated, Y_i^{sim} is the ith simulated value for the constituent being evaluated, Y^{mean} is the mean of observed data for the constituent being evaluated, and n is the total number of observations

Performance rating	$RSR = \left[\dfrac{\sqrt{\sum_{i=1}^{n} \left(Y_i^{obs} - Y_i^{sim}\right)^2}}{\sqrt{\sum_{i=1}^{n} \left(Y_i^{obs} - Y_i^{mean}\right)^2}} \right]$	$NSE = 1 - \left[\dfrac{\sqrt{\sum_{i=1}^{n} \left(Y_i^{obs} - Y_i^{mean}\right)^2}}{\sqrt{\sum_{i=1}^{n} \left(Y_i^{obs} - Y_i^{mean}\right)^2}} \right]$
Very good	$0.00 \leq RSR \leq 0.50$	$1.00 \leq NSE < 0.75$
Good	$0.50 < RSR \leq 0.60$	$0.75 \leq NSE < 0.65$
Satisfactory	$0.60 < RSR \leq 0.70$	$0.65 \leq NSE < 0.5$
Unsatisfactory	$RSR > 0.70$	$NSE \leq 0.5$

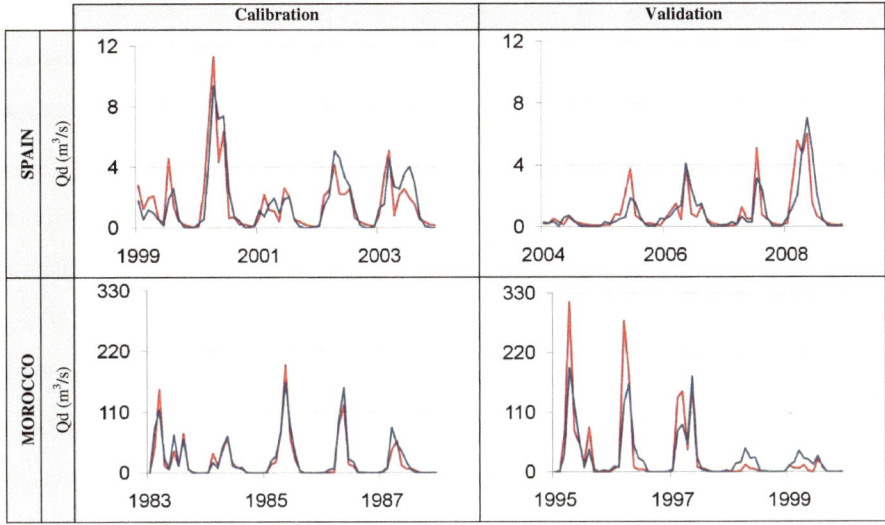

Fig. 3.2 a (*left*) Calibration results at monthly step. Shown are measured (*grey line*) and simulated (*black dashed line*) average daily stream flows (Qd, m³/s) in the Guadiaro river, Andalusia (6030 g.s.) and BOEM, Morocco (5764 g.s.). **b** (*right*) Idem for validation results

Following calibration of the SWAT model and the above-described validation processes, climate projections for the period 1961–2070 were introduced into the model in order to identify and quantify projected impacts on regional water resources. At this point in the assessment, two additional assumptions were required: firstly, land use cover was assumed to be constant throughout the 21st century; and secondly, reservoirs were simulated as uncontrolled.

3.4 Socio-economic Assessment

On the socio-economic side, the exposure of the IBRM communities under different Climate change and water management scenarios was analysed by local experts in stakeholder meetings and focus groups.

Two stakeholder meetings were held in May 2011 in order to identify the most sensitive water-dependent geographical areas and sectors in the IBRM. The first meeting was on the Spanish side of the IBRM in Malaga on 10th May 2011 where 6 stakeholders participated (Annex A specifies the institutions that participated in each meeting). The second meeting was held on the Moroccan side of the IBRM in Chefchaouen on 12th May 2011 with the participation of 9 stakeholders. Stakeholders were selected with the intention of combining different spheres of knowledge, experience, and sectors, including academia, public administration, the private sector and NGOs.

Table 3.3 Calibration and validation results. Shown are monthly means of stream flow values (Qm) from both simulated (Sim.) and measured (Meas.) data, and statistics, based on an illustrative sample of gauging stations

		Calibration			Statistics		Validation			Statistics	
		Period	Sim. Qm (m³/s)	Meas. Qm (m³/s)	NSE	RSR	Period	Sim. Qm (m³/s)	Meas. Qm (m³/s)	NSE	RSR
Spain	Guadiaro, 6030	1999–2004	1.7	1.6	0.7	0.5	2004–2009	0.9	0.9	0.6	0.6
	Palmones, 6083	1994–1999	0.3	0.4	0.8	0.4	2004–2009	0.4	0.3	0.6	0.7
	Guadalhorce, 6011	1999–2004	1.5	1.4	0.5	0.7	2004–2009	0.9	0.9	0.5	0.7
Morocco	M'douar, 612/04	1985–1990	14.8	12.9	0.5	0.7	1995–2000	13.0	16.0	0.4	0.7
	BOEM, 5764	1983–1988	24.7	21.7	0.9	0.3	1995–2000	18.6	18.1	0.8	0.5
	Khrofa, 1373/03	1981–1986	6.1	5.7	0.4	0.8	1995–2000	6.2	8.0	0.4	0.8

The discussions of the stakeholder meetings included the identification of at-risk areas and sectors within the IBRM. During the meeting, experts outlined their state of knowledge on the exposure of rural communities to global change effects and especially droughts, the sensitivity of natural resources, and the adaptive capacity of the IBRM facing future changes. Based on this information, future hydro-climatic and water management scenarios were drawn up to be used as starting points for focus group discussions to be held the following year (February 2012).

Following the methodology of Kallis (2010), the scenarios probe how the situation may develop by 2040 with respect to water resources and socio-political relations. Scenarios were constructed using the output of the stakeholder meetings and existing data on the IBRM [e.g. climate projections facilitated by Bruggeman et al. (2010)]. Table 3.4 shows the scenarios proposed for the IBRM.

These scenarios were presented in two focus group meetings held in February 2012. The objectives of the focus groups were to discuss the implications of the scenarios for water availability, livelihoods and human security in the region, to

Table 3.4 Scenarios proposed for the IBRM case study

			Hydro-climatic scenarios	
			Low impact scenario	High impact scenario
			Similar annual rainfall but with changes in seasonal precipitation patterns (less rain in summer and autumn) Slight increase of mean annual temperature	Significant decrease of annual rainfall (25%) Changes in seasonal precipitation patterns (less rain in summer and autumn) Significant increase of mean annual temperature (>2 °C)
Water management scenarios	Resource adaptation	Equitable distribution of water resources	Scenario A	Scenario B
	Water use prioritization	Prioritization of water use for a specific economic sector (tourism in Andalusia and unsustainable agriculture in Morocco)	Scenario C	Scenario D

identify vulnerable groups, and to define possible adaptation measures in order to ensure hydro-security in the long term. The first focus group was on the Spanish side of the IBRM in Malaga on 13th February 2012, where 7 stakeholders participated. The second meeting was held on the Moroccan side of the IBRM in Tetuan on 15th February 2012 with 9 participants. Although the same stakeholders and institutions were invited to the stakeholder meetings in May 2011 and the focus group meetings in February 2012, the participants were not identical (crossover of three institutions and one person on the Spanish side and crossover of four institutions and five people on the Moroccan side). The focus groups had a moderator who introduced and facilitated the sessions using different participatory tools. Particular attention was paid to consider all viewpoints, to discuss each of them, and to arrive at consensus. In both focus group meetings, the first two scenarios analysed were the extremes, scenarios A and D, where long discussions took place. Due to this, a short amount of time was left for intermediate scenarios, and the conclusions of scenarios B and C were less extensive than scenarios A and D. In the Andalusian focus group, most of the discussions centred on the effects of hydro-climatic scenarios. The water use prioritization scenario led to a heated debate due to the current lack of consistency in the way in which the law in Andalusia is applied when allocating water to different economic sectors. Water allocation was often felt to benefit the touristic sector (over agriculture for example). In the Moroccan focus group, participants tried to structure the dialogue by analysing economic sectors separately, but this approach predominated only in the beginning of the session. Afterwards, the discussion was organised according to scenarios rather than sectors.

In general, on both sides of the Reserve, Scenario A was perceived to present some modest benefits for all economic sectors, assuming the correct implementation of policies and control mechanisms for ensuring planned water distribution. This scenario provides the opportunity to test different adaptive measures in a low change scenario, which could then be escalated if climatic conditions worsen. Scenario B was qualified as the most probable future scenario; in this scenario the public administration would need to redouble its efforts so that policy measures ensure water security, promote social awareness, and convene participatory processes. The likelihood of water use scenarios C or D was initially questioned since both countries currently have regulations controlling water distribution, and water use prioritisation schemes may not be in place. However, during the discussion some participants argued that water use prioritisation already exists within the Reserve, but is a concern since existing schemes may result in unequal distribution. Scenario C was considered to be an extension of the current situation, due to lack of enforcement of existing schemes. Finally, Scenario D was taken to imply the rise of conflicts among water users and important losses in biodiversity, natural ecosystems, and cultural diversity of the region, whereas the protection of these was precisely the motive of creation of the IBRM.

3.4.1 Spanish Focus Group Outputs

In the following section the discussions of the Spanish focus group in relation to each scenario are summarised.

Scenario A involves a decrease in water resource availability, though this will be modest and will occur over a concentrated period of time. In general this scenario produces a large number of uncertainties since the benefits are counterbalanced by some negative effects and the final outcome will depend to a great extent on whether policies are able to quickly react to these changes.

Slight increases in mean temperature are expected to favour the time extension of the tourism season, mainly in mountain and rural areas. However, if current consumer preferences for coastal tourism continue, it is highly likely that there will be an economic migration of rural populations to these areas leading to a loss of local rural knowledge and identity. The effects of a longer dry summer could counteract the benefits of a longer touristic season. Decreases in water availability for touristic uses could affect the supply for resorts to maintain golf courses and swimming pools, which together with increased attractiveness of other more northern European regions, could negatively impact the sector.

Changes in seasonal precipitation patterns will increase the risk of flash floods, threatening populations located along river shores. Slightly higher mean temperatures will increase comfort within the IBRM and will reduce the consumption of energy for heating systems, but it could be counterbalanced by an increase in cooling requirements during the summer period as average summer temperature is expected to increase too.

A higher temperature will lead to a higher crop actual evapotranspiration, which together with lower precipitation in summer will reduce crop production and will lead to changes in cultivated crop types and varieties and their use. Water availability under this scenario is expected to support existing ecosystems and their corresponding natural value and services. However, higher temperatures will affect the cork production that is a major economic activity in the region, damaging therefore the forestry sector.

There are opportunities for research and implementation of adaptive measures in this scenario. The efficiency of measures can be assessed and validated under the new climatic conditions, and these measures could be strengthened if the climatic conditions worsen. Some of the proposed measures included basin reforestation plans, fluvial restoration, and development of infrastructures which are more resistant to flash floods, among others.

Scenario B produced similar concerns to Scenario A for human security, but in this case with a greater impact. In this scenario, lower precipitation will cause a significant decrease in available resources, forcing the development of water adaptation strategies. Government would need to promote public awareness about the scarcity of water resources and support the increased use of unconventional techniques such as water reuse and desalinisation in parallel to improvement of infrastructure efficiency and reduction in consumption. In this context, the promotion of citizen co-responsibility for water management can be very helpful for the governance of water resources, but a strengthening of administration to lead the process is required. Adaptive measures would need to ensure the maintenance of the sources of income for the rural population (agriculture and the promotion of sustainable tourism) and the protection of natural resources.

Scenario C assumed that economic developments will follow current trends where tourism is promoted as a main economic sector and consequently, the management of water supply will prioritise tourism, to the disadvantage of other rural activities or uses such as agriculture and the conservation of natural areas. This water use scenario is considered to be an extension of the current situation within the Reserve. In previous drought periods, the population of the IBRM suffered from water shortages, mainly affecting farmers, while coastal populations were able to meet their water demands by making new wells and diverting water from the Reserve. This scenario would produce inequity and force rural migration to the coast, leading to the loss of local knowledge, cultural and natural values in the Reserve.

Scenario D will cause a significant decrease in available resources and increased inequity between rural and coastal populations. Coastal tourism will benefit from higher water supply and temperatures, increasing economic benefits on the coast. The promotion of this tourism would lead to user conflicts with rural populations, reducing water availability and increasing rural migration to the coast. Abandonment of rural areas will lead to the loss of local knowledge and traditional activities within the Reserve.

The increased risk of flash floods will force the development of new infrastructures, with negative impacts on ecosystems. Water scarcity will increase awareness of the population, favouring the reduction of demand, the questioning of unsustainable infrastructures (e.g. golf courses) and positive perceptions of sustainable water management practices (e.g. use of wastewater). Moreover, water scarcity will stimulate the development of new and more efficient water technologies. Lower water availability will negatively impact water quality, increasing spending on water purification measures.

Agriculture will have to cope with lower water resources, while at the same time addressing the challenge of ensuring food security for local population as well as for tourism. Arboriculture (olives, chestnuts) and cork oak production will be reduced due to their increased water demands, whereas agriculture using plastic infrastructures will spread. Agricultural water demand will be reduced by developing more rational and efficient technologies.

Biodiversity and natural ecosystems will suffer from ecological flow shortages and eutrophication of water reservoirs. Higher temperatures will increase fire risk and the spread of diseases in the forests. Alien species will spread in the IBRM under warmer environmental conditions. Changes in land use, mainly urbanisation on the coast line, will modify the landscape, reducing the attractiveness of tourism in these areas.

3.4.2 Moroccan Focus Group Outputs

In the following section the discussions of the Moroccan focus group in relation to each scenario are summarised.

Scenario A would lead to a moderate decrease in available water resources over a concentrated period of time. Changes to rainfall patterns are perceived as more problematic than lower precipitation since this could imply an increase in flood events. In relation to water availability, an increase in saltwater intrusion into groundwater and a decrease in groundwater availability are indicated. Effects on ecosystem function are expected including forest decline events, as well as a change in migration patterns of fauna, shifting vegetation, and an increase in forest fire risk.

A temperature increase of less than 1 °C is not perceived as a serious problem since adaptation measures are able to cope with this level of change. This scenario of slight climate change could favour the extension of the tourism season, mainly in mountainous and rural areas. The scenario would likely lead to increased societal (professional and civil society) awareness about climate change and the need to implement adaptive strategies. In this sense, the scenario could be an opportunity to pursue more efficient water use. As a result there may be potential to promote a return to local knowledge and traditional uses of the environment in the Reserve. In parallel, an increased demand for agricultural products for export, partly to EU countries is expected.

Scenario B will cause a significant decrease in available resources. The more severe impacts on human security and water availability could be addressed through the implementation of water adaptation strategies in all economic sectors. These strategies must ensure food security for the local population, rural development, and equitable distribution of positive and negative externalities.

In recent decades strong economic competition between illegal crops (primarily cannabis) and alternative crops has emerged. Under *Scenario C*, the cultivation of illegal cannabis crops is expected to increase. In general, the growth of cannabis, while economically lucrative, is a form of maladaptation that actually worsens the effects of climate change. Cannabis cultivation in the IBRM is associated with various environmental problems (decrease of soil quality, soil erosion) because the crops are cultivated on high gradient slopes. However, more research is required to improve knowledge about the impacts of cannabis cultivation. This scenario would produce particular effects among the rural population in the short term since it implies an abandonment of traditional crops and their replacement by cannabis. In the mid-term, an increase in rural abandonment due to land degradation is expected in this scenario.

Under *Scenario D*, an increase in coastal population is expected as a result of water shortages in rural areas. Lower economic activity is expected to lead to a decrease in air pollution in certain areas. There will be a notable impact on bio-diversity with a significant loss of vulnerable species as the new climatic conditions are expected to favour the dominance of more adapted species as well as the invasion of new alien species. The salt production sector could be favoured by the new climatic conditions while land degradation and soil erosion are expected to worsen. This scenario could produce consequences for food security due to food shortages, which are also a key driver of migration amongst the local population.

3.5 Quantification of Impacts

On the basis of the focus group discussions, two categories of impacts on the IBRM were identified. These were (1) the impacts of climate change on the physical environment, including impacts on hydrology, climate and natural systems, and (2) impacts of climate change effects on socio-economic systems, including the agricultural and tourism sectors. Using the results of the socio-economic and hydro-ecological assessments described above, the following section explains these key impacts in detail.

3.5.1 Impacts of Climate Change on the Physical Environment

Impacts on hydrology and climate

Future scenarios show a high level of uncertainty about the degree of future climate change, however the primary concern is the impact of more extreme conditions in the IBRM; most specifically the effects of these conditions on water quality and quantity. In this assessment, stakeholders considered that there will be high demand for freshwater services in the IBRM and neighbouring areas, and that this is likely to continue increasing in the future as a result of continuing socioeconomic and demographic growth and the associated urban development. The results of our hydrological model for the region provide a quantitative basis for trends perceived by stakeholders.

According to future climate projections, the IBRM will experience increased aridity due to lower precipitation levels and higher annual mean temperature. The expected change in the water deficit index (WDI), which relates water availability to potential evapotranspiration (PET), is shown in Fig. 3.3. When this index is negative, it indicates that available water for vegetation requirements will be limited. The baseline period (Fig. 3.3a) shows a deficit in water availability across the whole IBRM; here, the deficit gradient decreases with increasing altitude. In the baseline scenario, coastal areas and other areas at low altitudes are subjected to greater pressure than mountainous or places with a more humid climate such as the Spanish Grazalema (having 2000 mm annual precipitation). In the future, increased aridity in coastal and low-altitude areas is expected, while no changes are expected in mountainous areas (Fig. 3.3b). If these trends proceed as described, this could lead to losses of climatically-suitable zones for a number of species (Thuiller et al. 2005;

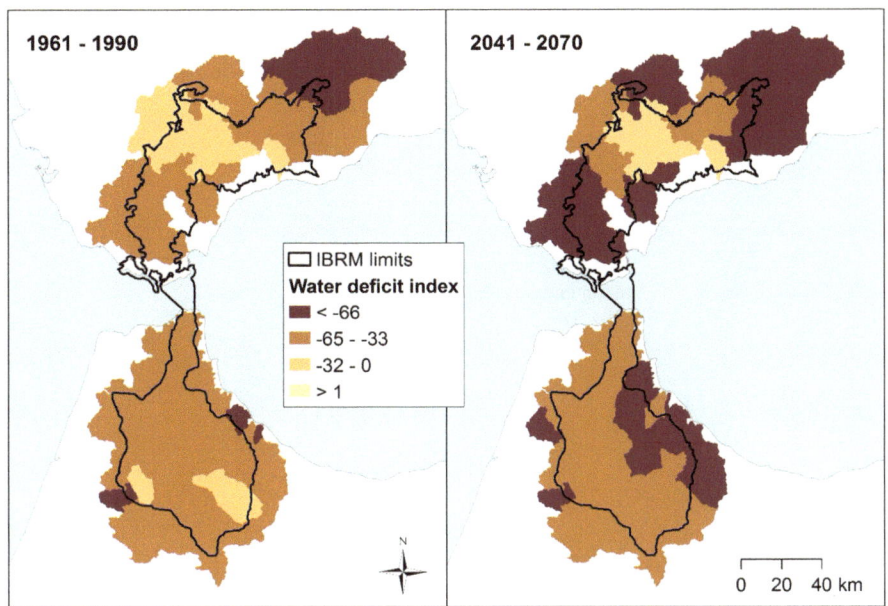

Fig. 3.3 a (*left*) Water deficit index (WDI, %) in the IBRM for the baseline (1961–1990). **b** (*right*) Idem for the 2041–2070 period. WDI = ((P-PET)/PET)*100; where P = Precipitation, PET = Potential evapotranspiration; negative index values indicate limitations in water required to cover vegetation requirements

Serra-Diaz et al. 2012), thereby altering species distribution. Also, areas at lower altitudes have the highest population densities. These observations indicate a future spatial segregation of ecological and social impacts. As such, while changes in species composition may be relevant for biodiversity concerns, the impact on human activities will be limited in higher altitude areas. On the other hand, there will be a greater pressure on water demand for a multitude of uses as well as a general water deficit at lower altitudes and along coastal areas.

According to the hydrological model, the anticipated mean annual temperature increase of 2.4–2.5 °C will contribute to an increase in potential evapotranspiration (PET) of approximately 8.6–9.6%, leading to an increase in the water demand of vegetation (Fig. 3.4; Table 3.5). Although PET increases with temperature, the actual evapotranspiration (AET) is expected to decrease between 4.5 and 7.4% since this depends on precipitation and temperature. This decrease in AET indicates that vegetation water demands will not be met, negatively affecting the development of vegetation in the IBRM.

Using hydrological simulations and climate projections, a generalised reduction in stream flows is predicted over the 21st century. A reduction of between 28.1 and 30.3% in stream flow is predicted by 2070 (Fig. 3.5), with the greatest reductions expected on the Spanish Atlantic Slope. In addition, an increased seasonality in stream flows can be expected. Also, the probability of extreme events such as flash

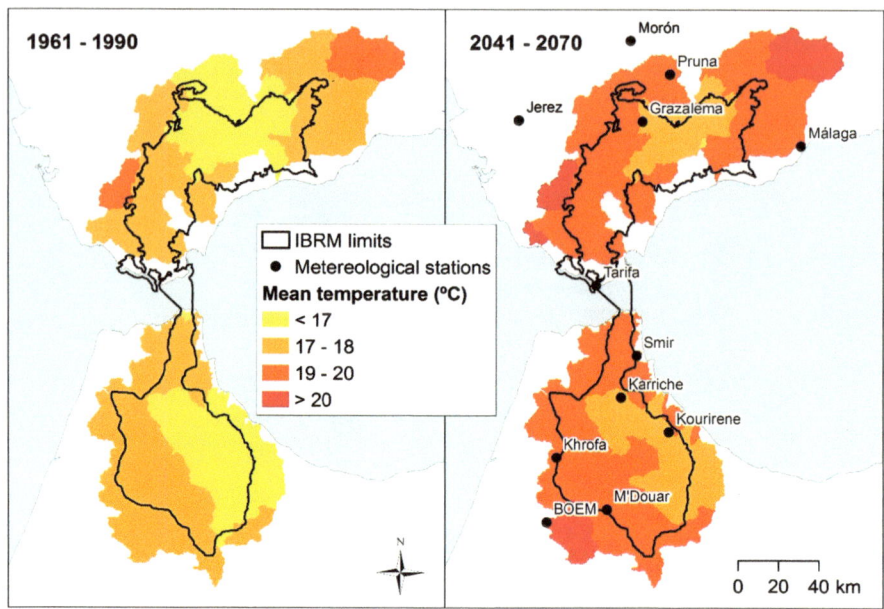

Fig. 3.4 Mean temperature (°C) in IBRM subbasins for the baseline time period 1961–1990 (*left*) and the 2041–2070 period (*right*)

Table 3.5 Expected change in mean annual precipitation (P) and temperature (T), potential and actual evapotranspiration (PET and AET, respectively) and daily stream flow (Qd) in the IBRM comparing the 2041–2070 period with the baseline

	P (%)	T (°C)	PET (%)	AET (%)	Qd (%)
Andalusia (Spain)	−17.6	+2.5	+9.6	−7.4	−30.3
Morocco	−14.3	+2.4	+8.6	−4.5	−28.1

flooding will increase, with a high risk of soil erosion as a result of these events (Moore et al. 1998; Agoumi 2003; Ruiz Sinoga et al. 2010). Subsequently, a general reduction on groundwater recharge is also predicted in this area (CEDEX 2012).

From an ecological standpoint, the expected reductions in streamflow will imply longer periods of low stream flow affecting the environmental quality of streams. With this outlook, riparian ecosystems in some parts of the Reserve would be profoundly affected. Also, decreases in soil water availability and changes in intra-annual variability could impact the development of woodlands, the suitability of the environmental conditions for certain forest species, and the supply of water for agricultural and urban uses (Schröter et al. 2005).

Fig. 3.5 Projected changes in stream flow (%) in the IBRM when comparing the 2041–2070 period with the baseline (1961–1990)

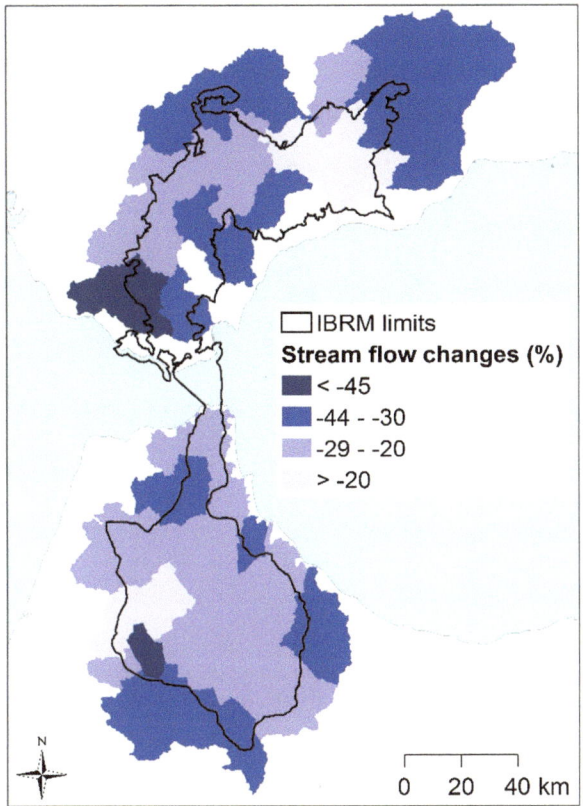

Impacts on natural systems

Stakeholders were concerned about the impact of increased environmental aridity on forests and natural ecosystem processes, as these impacts are likely to result in losses in ecosystem integrity and biodiversity. The two most relevant impacts to be expected in the IBRM are forest decline (Jump et al. 2006; Sarris et al. 2007) and increase in forest fire events (Moriondo et al. 2006). The protection of forest areas and shrubland was one of the primary motives for the creation of the Reserve, these being the most extensive and representative natural systems of the IBRM.

The amount of water stored in the soil is an integral parameter for forest health status, forest sensitivity to mortality, and fire risk. Changes in soil water storage are used to assess climate change impacts on natural systems (Fig. 3.6). According to the model, forested areas may suffer a greater reduction in soil water storage than shrublands. This indicates a higher potential vulnerability of forests in the IBRM to future climate trends. The hydrological model also indicates that forests on the Moroccan side of the IBRM in particular may be more sensitive to future conditions than on the Spanish side. Since shrublands are better adapted to the increasingly arid conditions, their distribution and their extension in the IBRM may expand.

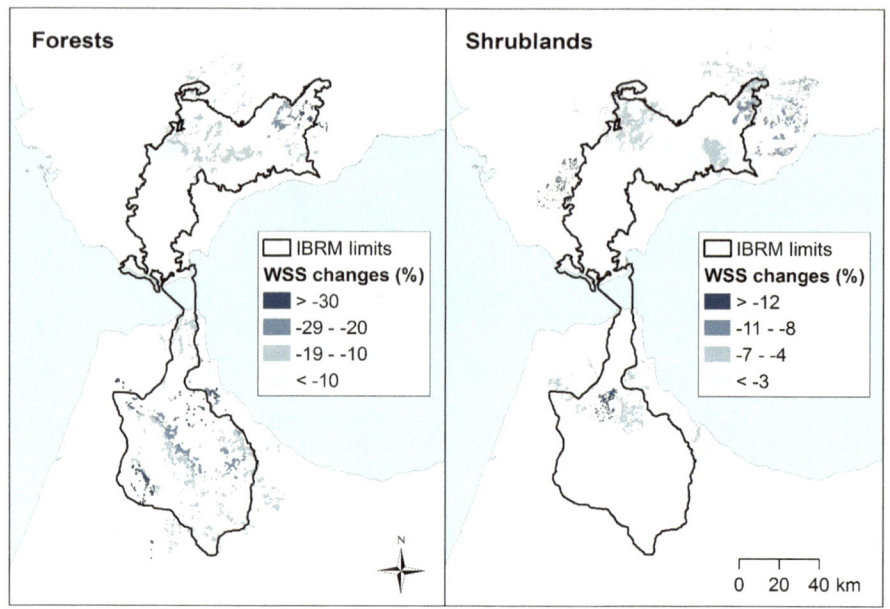

Fig. 3.6 a (*left*) Changes in water stored in soils (WSS) (%) in IBRM forests comparing the 2041–2070 period with the baseline (1961–1990). Forest areas are circled with the background colour to make them more visible. **b** (*right*) Idem for shrublands

3.5.2   Impacts of Climate Change on the Socio-economic Sectors

Impacts on the agricultural sector

Agriculture, including livestock, has been identified as the most exposed and economically-sensitive sector to climate change effects; this is particularly the case on the Moroccan side of the IBRM. The greatest risk to subsistence farming systems is unfavourable weather, and this is especially true on marginal land where drought years can directly affect a farmer's income. Traditional agriculture and extensive farming systems are still the main means of subsistence for a large part of the Moroccan population in the Reserve. In Andalusia, recurrent droughts have led to serious losses and damages to arable and pastoral sectors, though the affected surface area is generally small. In general, the most affected crop types in the region are vineyards and olive trees.

The hydro-ecological assessment shows a general future increase in water demand for crops and pastures (Fig. 3.7). Water demand includes the supplementary water needed by crops and pastures to maintain current production standards, and is gauged based on the water stored in soil. Here, irrigated and rainfed crops are considered together in order to homogenise the analysis.

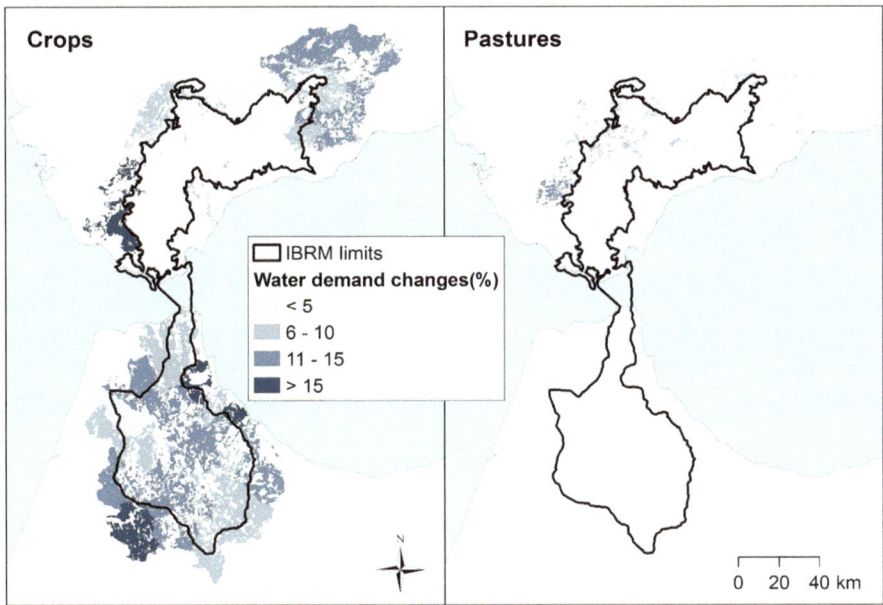

Fig. 3.7 a (*left*) Crop water demand changes (%) in IBRM crop areas comparing the 2041–2070 period with the baseline (1961–1990). **b** (*right*) Idem for pasture areas

The results show important differences in crop patterns when comparing the results for Spain and Morocco, which require different management decisions. In the Spanish part of the IBRM, low crop activity within the protected area strongly contrasts with significant agricultural pressure in neighbouring areas bordering the IBRM. As shown in Fig. 3.7a, the projected increase in water demand will not have significant effects within the Reserve in the 21st century. On the Moroccan side, it is fairly evident that there is no significant difference in the density of crops inside or outside the IBRM. Nevertheless, a higher demand for water resources is predicted in the surrounding areas (Fig. 3.7a) where 7.5–16% increase in water for irrigation will be necessary to maintain current crop production standards up to the year 2070. Pastures are spread throughout the Spanish side of the IBRM (Fig. 3.7a). Increased water demand is predicted outside the Reserve, while pastures inside the IBRM are rainfed, so no important change in water demand is predicted.

Impacts on the tourism sector

In the assessment, the tourism sector was identified as being very sensitive to climate-related impacts, especially during the summer period. Stakeholders were concerned about effects of higher temperatures on the suitability of the region for tourism, as well as the impact of likely water shortages.

The hydro-ecological assessment includes the analysis of the impacts of future increased temperatures according to several comfort indicators. These are the number of: hot days (Tmax > 30 °C); very hot days (Tmax > 35 °C); and tropical nights (Tmin > 21 °C). For Malaga, an Andalusian coastal city, the results show an increase of 39 hot days, 46 tropical nights, and a slight increase in very hot days (4) per year by the year 2070 when compared to the baseline years (1983–2009).

On the Spanish side of the IBRM, the projected changes in climatic conditions will result in a decrease in the Tourist Climatic Index[1] over a 50–60 year timeframe (Ciscar et al. 2009). The changes will have a major impact on climate during the summer period, and as a consequence it is expected that tourism will be diverted to other parts of Europe during this season. While this may have the positive effect of decreasing demand for water, it will have a negative impact on the economy. These considerations highlight the need for a long-term strategy as far as tourism is concerned. No relevant information is available for the Moroccan IBRM, but similar impacts could be expected due to the similar nature of the projected climatic change.

3.6 Conclusions

At present, most global change impact studies have been undertaken for distinct sectors. In the future, integrated projects need to be conducted in order to obtain a global overview of cumulative global change effects (IPCC 2007a). This assessment combines a hydro-ecological impact assessment with an assessment of the social dimensions of water use and management.

Future climate scenarios paint a picture of a more arid climate and strong changes in water dynamics within the IBRM over the 21st century. An increase in mean annual temperature of two degrees is predicted, along with higher water stress and scarcity within the region. Similar patterns within the same order of magnitude have been documented in other regional projections (Moreira and Ribalaygua 2007; AEMET 2009). Altitude appears to have an effect on predicted patterns, corresponding to higher aridity in coastal and low-altitude zones hosting the highest population densities, whereas no change is predicted in mountainous areas: this points to a clear spatial segregation of ecological and social impacts of climate change. Specifically, at higher altitudes some impacts on species composition relevant for biodiversity may occur, but there is low relevance for human activities. In contrast, at lower altitudes and along coastal areas there will be increased pressure on water resources, increased demand for different uses, and a larger water deficit.

The predicted changes in precipitation will have direct impacts on streamflow in the Reserve, estimated to be reduced by 30%. Similar reductions have been observed in other parts of the Mediterranean, as discussed by Senatore et al. (2011) and Pla et al. (2012). As a consequence of this, the seasonality of water in

---

[1]The tourist climatic index is based on the notion of 'human comfort' and consists of a weighted index of maximum and mean daily temperature, humidity, precipitation, sunshine and wind.

streams will increase, with important effects on the water supply which in turn is subject to competition from multiple uses including those which are environmental, agricultural or urban (Schröter et al. 2005). As a result of the expected changes in water availability, the conservation of environmental quality in riparian ecosystems and streams will be highly jeopardized. Indeed, some water bodies in the area are already at risk of not meeting the environmental objectives of the European Water Framework Directive (2000/60/EC) (Iglesias et al. 2007).

Increases in temperatures during the 21st century are expected to impact natural and agricultural ecosystems through effects on growth and development of vegetation (Vicente-Serrano et al. 2013), vegetation phenology (Giannakopoulos et al. 2009) and species distribution (Serra-Diaz et al. 2012), leading to losses of climatically-suitable areas for a number of species. At the landscape level, there may be changes to crop species and varieties used, provoked by (vegetation) phenological changes as well as lower water availability, effectively modifying existing agronomical techniques and the agronomic calendar (Moriondo and Bindi 2007; Giannakopoulos et al. 2009).

An increase in aridity will have a direct impact on forest ecosystems and natural processes, likely causing a loss in ecosystem integrity and biodiversity. The assessment demonstrates greater losses in water stored in forest soils than in shrubland areas, pointing to the higher potential vulnerability of forests to future climatic trends. The range of shrublands might expand since they are more adapted to arid conditions. The two most important expected impacts are forest decline (Sarris et al. 2007) and increase in forest fire risk (Moriondo et al. 2006). Meanwhile, lower soil water availability and changes in intra-annual variability may impact the development of woodland and the suitability of the environmental conditions for the development of some species (Schröter et al. 2005). Predicted climatic changes include an increase in hot days, tropical nights and water shortages leading to higher risks to human health, including higher risk of mortality. Levels of tourism in the IBRM are likely to be affected during the hottest summer months.

Under the more moderate scenarios, cooperation among sectors is facilitated, which favours the maintenance of local knowledge and traditional uses of the Reserve. However, it is evident that in Andalusia the positive results of this scenario are likely to be outweighed by the negative impacts. On the Moroccan side of the IBRM, the most undesirable scenarios lead to food shortages, soil degradation associated with unsustainable agriculture, and human migration. In Andalusia, the most severe scenarios would negatively impact the rural population, cause migration towards the coast, and result in a loss of rural identity, values, and knowledge. In both countries, the conservation of natural and cultural diversity—which was precisely the rationale for the creation of the IBRM protected area—would be threatened.

The results of the impact assessment underscore the substantial impact of regional climate change on water resources. They also demonstrate the importance of incorporating these types of analyses into adaptive management in the Mediterranean region. The next great challenge will be the implementation of policy frameworks and management systems for water resources so that the sustainability of natural systems and human security can be safeguarded. Furthermore, since socio-economic activities within the Reserve (agriculture and rural tourism) are in strong competition for water resources with neighbouring areas (coastal tourism and some major urban and industrial areas), activities within the Reserve require special support and attention so that they may be sustained in the future.

Chapter 4
Policy Assessment

Abstract Adaptation to climate change requires collective action and while this can be self-organised, governance mechanisms and institutions play a very important role in enabling such collective action (Adger 2003). By contrast, a lack of official policies and frameworks may impede adaptation or increase the vulnerability of certain groups (Madzwamuse 2010). This chapter provides an extensive overview of policy frameworks relevant to mitigating water-related risks to human security in the Intercontinental Biosphere Reserve of the Mediterranean (IBRM) situated between Morocco and Spain. It assesses their effectiveness in reducing the Reserve's vulnerability to climate change i.e. the adaptive capacity of political systems in the Reserve. The chapter covers policies, strategies, laws and governance mechanisms in the areas of climate change adaptation, water resource management, disaster risk reduction, forests and natural systems, agriculture, tourism and human development. Policy frameworks at national level in Morocco and Spain as well as at regional and local level are included. To complement this primarily desk-based overview, insights are brought in from stakeholders and experts on the challenges and opportunities of this policy framework for addressing climate-induced risks to human security in the IBRM.

Keywords Mediterranean basin · Spain · Morocco · Climate change · Climate adaptation · Adaptive capacity · Human security · Food security · Water security · Floods · Water scarcity · Drought · Tourism · Agriculture · Policy frameworks

4.1 Approach and Methodology

This chapter provides an overview of the current policy framework governing the IBRM, on the basis of policies and programmes at national level in Morocco and Spain as well as at regional and local level. To complement this overview, insights from stakeholders and experts are brought into review some of the challenges and opportunities that this policy framework presents for addressing climate-induced risks to human security in the IBRM. The analysis focuses on policies that address

© The Author(s) 2017
D. Abdul Malak et al., *Adapting to Climate Change*,
SpringerBriefs in Environmental Science, DOI 10.1007/978-3-319-51680-6_4

the key areas identified by the impact assessment in Chap. 3, i.e. policies directly tackling climate change, water quality, availability and use, as well as sectoral policies for agriculture, natural systems and tourism.

The overview of the policy frameworks comes from research carried out in a primarily desk-based exercise, complemented with additional information through stakeholder interviews. The policy framework in the IBRM is assessed for its effectiveness on the basis of stakeholder perceptions informed by two stakeholder focus groups, which were carried out in Malaga, Spain on 13th February 2012 and in Tetuan, Morocco on 15th February 2012. In the following sections, the rationale for the analysis of policy frameworks is outlined (Sect. 4.2), looking in turn at national-level in Spain (Sect. 4.3) and in Morocco (Sect. 4.4), and the local level of the IBRM (Sect. 4.5). Drawing on insights from the focus groups and interviews carried out in the context of this and the research in Gerstetter et al. (2012), stakeholder perceptions and assessments of the policy frameworks are detailed (Sect. 4.6) and the capacity of these frameworks to address adaptation to climate related risks to human security are discussed.

4.2   Policy Frameworks

In the analysis of policies relevant to the IBRM, this chapter follows the approach adopted by Gerstetter et al. (2012) and focuses on 'policy frameworks'. Terms such as "governance", "political factors" or "political system" may help to explain *why* policies are or are not adopted, but do not allow for a definition and analysis of the actual policies and laws that govern the IBRM. Policies can be understood as different types of overarching action undertaken or supported by governments or public actors such as international organisations (e.g. laws, strategic programs or long-term and consistent approaches on how to deal with certain issue). However, the term 'policy framework' is used here, as it is somewhat broader than policies in that it seeks to capture if/how in a given polity different individual policies (e.g. a water law or an adaptation strategy) are adopted as well as if/how they are coordinated or integrated. By taking this approach, it is hoped that a more comprehensive assessment can be made of the extent to which climate-induced risks to human security are addressed by the responsible political institutions in the IBRM.

Adaptation policies can enhance human security e.g. by improving the socioeconomic situation of certain population groups and thus their capacity to adapt to changing conditions, by improving the state of the natural environment, thus limiting the impact of climate and change, or by protecting populations against specific risks (e.g. floods) that endanger human security. For this reason, the policy frameworks analysed in this chapter cover not only those explicitly targeting adaptation to climate change, but also those which address climate impacts in a cross-cutting way, such as agricultural policies targeting drought or water policies targeting impacts of flooding on water quality. Despite the focus here being on 'top-down' adaptation approaches that typically rely on modelling of climate change impacts to identify suitable interventions, rather than on 'bottom-up' approaches (Dessai and Hulme 2004), this does not mean that adaptation efforts of other policy actors, including the bottom-up, more ad hoc decisions that individual or communities take in response to changing environments are not equally important for overall adaptation efforts.

4.3   Spanish Policy Framework

This section provides a broad overview of policies and programmes in Spain that address water related impacts of climate change and adaptation highlighting their consideration or impact on human security and vulnerability where possible. Due to the structure of the Spanish policy framework, policies from EU, National and Autonomous Community level are described.

4.3.1   Adaptation to Climate Change

The Spanish government was one of the first European countries to pass a National Adaptation Strategy (NAS) to climate change, establishing the National Climate Change Adaptation Plan (PNACC) in 2006. The PNACC's general objective is to evaluate the vulnerability of different socio-economic sectors and environmental systems that are sensitive to climate change, and to study the adaptation options for these sectors and systems. Quite unusually, the PNACC explicitly addresses the human security dimension of climate change. It states that in the evaluation of climate change impacts and vulnerability for each sector or system, human security impacts and other human security-related aspects like food security, poverty or social inequality must be considered—along with factors that determine adaptation options related to these impacts. Institutions, governance, demography, culture and social values are stated as examples. However, the PNACC remains imprecise in this field and does not provide for further descriptions how human security impacts will be addressed and how related adaptation options can be developed. To further define priority areas and actions, the PNACC produces work programmes. Each programme is intended to build on the knowledge gained in the previous step, ensuring a certain level of flexibility. Water resources are consistently identified as a priority system to be evaluated and addressed due to water's cross-cutting impact on other sectors (e.g. agriculture or tourism). Some results in integrating climate change adaptation into different sectoral policies have so far been achieved in water, desertification and forests (all under control of the Ministry for Agriculture, Food and Environment, MAGRAMA). Climate change adaptation is now integrated into a range of legislative instruments, covering; for example, coastal zones,[1] flood risks[2] and desertification[3] as well as river basin plans (see Sect. 4.3.2). Up to date information on future climate change impacts in the short, medium and long-term must also be taken into account in Environmental Impact Assessments and Strategic Environmental Assessments.[4]

What is more, mainstreaming has also been successfully implemented beyond the environmental ministry of MAGRAMA. Sectoral integration of climate change adaptation in health policy was made in 2010 with the joint set-up of an Observatory on Health and Climate Change by the Ministries of Health (MSPS) and the MAGRAMA. This can be regarded as an important achievement of the PNACC towards further integration of climate change adaptation in sectoral policies, as this measure of integration goes beyond the competences of the MAGRAMA and implies both inter-departmental and inter-ministerial coordination.

Due to the decentralised political framework in Spain, the Autonomous Communities (CCAAs) play a major role in the definition and implementation of environmental policy. Most of the CCAAs have developed their own climate change adaptation plans or strategies, either as individual strategies or within general climate change programmes. For this reason, the PNACC sees coordination between the national level and the 17 Autonomous Communities as crucial in order to avoid duplications and ensure complementarity of the PNACC and the plans, strategies and adaptation programmes of the CCAAs. At the same time this coordination has also been identified as a major challenge (Swart et al. 2009). Indeed, the climate adaptation work programmes that implement the PNACC have focused on strengthening coordination between the administration at the national and autonomous levels, through for example the creation of an expert group (Grupo de Trabajo sobre Impactos y Adaptación, GTIA) and an exchange platform (Plataforma de intercambio de información sobre adaptación, AdapteCCa). The PNACC states within its aims to continuously provide assistance to all interested administrations and organizations, both public and private, in evaluating the impacts of climate change in the respective sector, facilitating evaluation knowledge, tools and methods. The ministry responsible—MAGRAMA—also highlights the public participation process that accompanied the formulation of the plan, although Swart et al. (2009) report that only around 50 enquiries were received in the public consultation process.

As an Autonomous Community of Spain, Andalusia is responsible for creating its own Climate Change Strategy (Estrategia Andaluza ante el Cambio Climático 2002) with support from its Monitoring Panel and Technical Secretariat that

---

[1]Ley 2/2013, de 29 de mayo, de Protección y uso sostenible del litoral y de modificación de la Ley 22/1988, de 28 de julio, de Costas.

[2]Real Decreto 903/2010, de 9 de julio, de Evaluación y Gestión de riesgos de inundación.

[3]ORDEN ARM/2444/2008, de 12 de agosto, por la que se aprueba el Programa de Acción Nacional de Lucha contra la Desertificación.

[4]Ley 21/2013, de 9 de diciembre, de evaluación ambiental.

involves the development of instruments for prevention and environmental control, implementation of specific policies for forestry and biodiversity and the development of planning instruments and monitoring indicators including the Information System of Environmental Climatology. In 2010, it passed the Andalusian Action Plan for Climate: Adaptation Programme (Plan Andaluz de Acción por el Clima: Programa de Adaptación) which is, interestingly under the responsibility of the Commission of Economic Affairs, with the Environmental Council in a supporting role. Indeed this emphasis is reiterated by the Adaptation Programme which points out that adaptation is not an 'environmental policy' but rather a horizontal policy with measures that should be promoted from different areas. The focus of the Programme is to develop sectoral measures; extend the knowledge base on climate impacts; increase adaptation efforts by the administration at CCAA and local level and the private sector; and support training and education.

4.3.2 Water Resources

In order to face water scarcity and growing demands, water policy in Spain has traditionally had a strong focus on increasing supply through infrastructural projects. However, the proposal for a large scale transfer under the National Hydrological Plan (NHP) in 2001 was a political failure (Box 4.1) and a new approach was sought. The AGUA Programme (Ley 11/2005, Actuaciones para la Gestión y la Utilización del Agua—Law 11/2005, Actions for Water Management and Use) aims to improve water management, availability and quality in Spain and has been described as a paradigm shift in Spanish water policy (Downward and Taylor 2007). The AGUA Programme introduced new objectives for Spanish water policy that address the requirements of the European Water Framework Directive (WFD) such as the inclusion of economic instruments and demand-side management measures. The Programme explicitly highlights its relevance in respect to climate change and its impacts on water resources in Spain: the development of water resources from new sources (i.e. desalination) is seen as the suitable measure to guarantee water supply "independently from the climatic situation". In Spain, desalination is seen as a way to alleviate conflicts and human security concerns over possible water use restrictions or the need for large-scale water transfers as proposed by the NHP.[5]

However, there is no explanation of how water tariffs, and specifically desalinated water tariffs will be affected by the full-cost-recovery principle. On the one hand, cost-coverage is a main goal, but on the other hand the policy mentions that water costs will remain in all cases at a minimum level. Furthermore, the AGUA Programme has been criticized for its lack of specific targets or indicators

---

[5]See the environmental sustainability report on the Urgent Actions of the AGUA Programme in the Mediterranean Basins (AUPACM), prepared by the Environmental Ministry for further details.

(environmental or economic) by which the successes and failures of the Programme can be assessed (Downward and Taylor 2007). It has also been argued that the AGUA Programme maintains a focus on "capital-intensive solutions, centred on supply management, which would ultimately benefit the traditional hydraulic policy community, with its strong preference for technical (capital-intensive) options" (Lopez-Gunn 2009).

Box 4.1. The National Hydrological Plan (NHP)
The National Hydrological Plan (NHP) of 2001 proposed the transfer of water from the Ebro river to drier provinces as a central measure. It was planned to transfer 820 hm^3 of water up to 750 km to the Júcar, Segura and Mediterranean basins of Andalusia (then named the Sur basins), and additionally 200 hm^3 were planned to be transferred to the Metropolitan area of Barcelona, located in the internal basins of Catalonia (Albiac et al. 2006). However, the NHP provoked strong disputes between the regions involved in the Ebro transfer: the governments of the Autonomous Communities of Aragón and Catalonia opposed the project claiming the area-of-origin rights; on the other side, southern regions that would have profited from the transfer, argued that inter-community basins are under national jurisdiction and water transfers between basins are projects of national interest (Garrido and Llamas, Water policy in Spain, 2009). Furthermore, the dispute over the Ebro transfer was not only a matter between the regional governments; many civil society initiatives, non-governmental organizations and the scientific community were strongly involved in criticizing the plan (Font and Subirats 2010). In addition to the strong opposition to the transfer within Spain, the European Union was reluctant to fund the project due to potential misalignment with the objectives of the new Water Framework Directive (WFD) which focused on cost recovery models and strongly backed the movement towards a "new water culture" in Spain (Tàbara and Ilhan 2008). This eventually took the form of the AGUA Programme which replaced the proposed NHP through Law 11/2005.

The WFD made river basin level water management a requirement for EU Member States. However, Spain has a long historical experience with this approach, having created its first basin authorities (Confederaciones Hidrográficas) in 1926. The composition and functions of the Spanish river basin authorities, are regulated by the Water Law of 1985 and the Royal Legislative Decree 1/2001 (Real Decreto Legislativo 1/2001). In 2005, the government also initiated a process to reform the Spanish water law, incorporating the principles of the AGUA Programme into national legislation. The AGUA Programme provides for a further reform of the basin authorities that incorporates regional governments in the decision making process and the control of water use and quality. It also aims to increase citizen participation in water management and the creation of public water banks in each basin, in order to allow the reallocation of historical water rights based on new criteria such as equity, efficiency and sustainability. Indeed, the Royal Decree 907/2007 Regulation for Hydrological Planning, modified by the Royal Decree 1161/2010 (Real Decreto 1161/2010, por el que se modifica el Real Decreto 907/2007, por el que se aprueba el Reglamento de la Planificación Hidrológica) provides for an increased level of active public participation in the planning and management of water resources.

As noted above, PNACC's objective is to integrate climate adaptation into sectoral policies and at different levels of governance. In the case of water, it is the implementation of the WFD that is seen to be the relevant process for this integration. The PNACC states that although the term "climate" does not appear in the text of the WFD itself, impacts of climate change on water quality and availability need to be considered. As the implementation of the WFD primarily rests upon action at the river basin level, the Spanish government has made it obligatory for hydrological planning to include the evaluation of climate change impacts through the Royal Decree 907/2007. Furthermore, the Orden ARM/2656/2008 requires each river basin management plan (RBMP) to evaluate future climate change impacts on water resource availability on the basis of scenarios provided by the MAGRAMA. In case own evaluations are not carried out in the planning process, the Orden provides the concrete percentages to be applied corresponding to each basin in terms of reduced water availability corresponding to the year 2027. The 2010 Autonomous Water Law of Andalusia (Ley 9/2010, de 30 de julio) guarantees the basic water needs of the population while at the same time ensuring that social and economic development of Andalusia are in keeping with the protection of aquatic and terrestrial ecosystems and sustainable use of water in the Autonomous Community. The Law also provides for the evaluation and management of drought and flood risk.

As in other semi-arid countries, droughts are a common feature of the Spanish climate. Llamas (2000) points out "when this physical fact is forgotten, economic and social tensions ensue." Spain has in the past suffered from the inability to make timely predictions of droughts and when these occurred they were often regarded as emergency situations that led to costly infrastructural emergency measures to ensure water supply (ibid.). These issues, coupled with the additional challenges in the light of climate change—a projected increase in the frequency and severity of droughts—made the need for a policy shift evident. The National Hydrological Plan 2001 (Law 10/2001) introduced new regulations for drought management in Spain that included: a global hydrological indicator system for inter-community basins for drought forecasting and alerts to be provided by the MAGRAMA; the elaboration of drought management plans (Planes especiales de sequía)—complementary to the river basin management plans—by the inter-community basin authorities; and drought emergency plans. Drought management plans (DMPs) were finally adopted in March 2007 with the objective of minimising environmental, economic and social impacts caused by droughts. Even though only obligatory for inter-community basins, the article explicitly stresses that measures can be adopted by CCAAs that are in charge of internal basins. As with RBMPs, the river basin authorities (Confederaciones Hidrográficas) are the institutions in charge of developing and implementing DMPs based on a public participation process which can help to ensure more transparent decision making, thus reducing the potential for conflict. Even though the guidance document for elaborating DMPs does not explicitly refer to climate change, the policy measure itself is highly relevant in respect of a changing climate. Ensuring human security is an implicit goal of the policy, as it aims to "sustain population life and health".

The experience and capacity that the Spanish government has built during the development of its DMPs has had a strong influence on the European process of drought policy development. The Drought Management Plan Report (COM 2007) gives general guidelines for the development of DMPs in accordance with the objectives of the Water Framework Directive and builds largely on the Spanish model. Through its experience in changing from an emergency approach to establishing a planned management approach in order to face droughts, Spain has gained a capacity that may act as a reference not only for other semi-arid countries in Europe, but for all Mediterranean countries.

The policy framework for flood prevention is, to a large extent, informed by the 2007 European Floods Directive. The Directive requires Member States to identify areas at risk of flooding, produce flood risk maps and management plans focused on prevention, protection and preparedness. These actions are to be coordinated with the river basin management plans of the European Water Framework Directive. In Spain, the MAGRAMA has set up Automatic Hydrological Information Systems (SAIHs) to assist with the prediction and management of floods and the National Cartographical System of Flood Zones which defines areas of potential inundation. The Real Decreto 903/201 that transposes the Floods Directive also reinforces the coordination of all administrative bodies in order to reduce damages taking into account not only structural measures, but also supporting management and land-use planning measures. In addition, the 2008 amended Public Water Resources Regulation (Real Decreto 9/2008 por el que se modifica el Reglamento del Dominio Público Hidráulico aprobado por el Real Decreto 849/1986) focuses specifically on the flood risk management and civil protection. In January 2016, the Spanish government approved 16 plans for flood risk management along with 800 million EUR of investment to apply more than 1000 measures until the year 2021. The creation of the plans was carried out jointly with the Spanish Office for Climate Change (Oficina Española de Cambio Climático) and for the first time, climate change was incorporated as a variable in such plans (EFE 2016).

4.3.3 Agriculture and Rural Populations

The key policy framework for agriculture in Spain is derived from the EU level. The Common Agricultural Policy (CAP) aims to strengthen the competitiveness and sustainability of EU agriculture by providing direct payments and market measures to farmers and finances programmes for the development of rural areas in the EU. For the period 2014–2020, the CAP will invest around 45 billion EUR in Spain's farming sector and rural areas (COM 2014). Despite this apparently high level of support, 41% of Spaniards find that the level of financial support through the CAP is too low (Eurobarometer 2014). At the European level, key political priorities under the CAP include jobs and growth, sustainability, modernisation, innovation and quality, although Spain is given flexibility to direct the funds to its own priorities. Since the 2013 reform of the CAP, 30% of direct payments are linked to environmentally-friendly farming practices: crop diversification, maintaining permanent grassland and conserving 5% of areas of ecological interest. These funds can provide an important safety net, particularly important in Spain where the primary sector provides 4.2% of jobs (COM 2014). In 2012, the EU spent around 632 million EUR on market measures in Spain (ibid). Box 4.2 provides an example of a rural development project supported by the CAP in the Andalusian region.

Box 4.2. Andalusia shepherds school [Source: Spain, Common Agricultural Policy Factsheet (COM 2014)]
This project aims to provide training to livestock farmers in order to promote extensive grazing practices, favour generation renewal, and increase the quality and added value of livestock products. Extensive grazing is a useful tool to prevent forest fires, preserve biodiversity and maintain landscapes. Training topics covered are animal welfare, feeding techniques, fire prevention practices, preservation of the environment, or management of livestock holdings. More than 30 shepherds have benefited from the four editions of the project. The total cost of the project was 230,000 EUR with an EU contribution of 172,000 EUR.

In accordance with the EU Regulation 1698/2005 on rural development, the Spanish government developed a National Strategic Plan covering its rural development policy that aims to be applied via Rural Development Programmes (one national and 17 at autonomous level). In their most recent iteration, these programmes focus on three key objectives: increasing agricultural competitivity; ensuring sustainable management of natural resources and climate action; reaching a balanced territorial development of rural economies and communities including the creation and conservation of jobs in the period 2014–2020. Each Autonomous Community identifies the priorities relevant for its territory. The Rural Development Programme for Andalusia 2007–2013 (Programa de Desarrollo Rural de Andalucía 2007–2013, PDR) seeks to increase the quality of life in rural areas and greater economic diversification in the region. For the period 2014–2020, the programme has not yet been fully adopted, but focal areas include the protection and improvement of the environment; improvement of quality of life; competitivity of small and medium enterprises; and research, innovation and ICT.

4.3.4 Tourism

Spain is a world leader in tourism: it is the world's third most popular tourist destination in terms of arrivals, it was 2nd in tourism earnings worldwide (and 1st in Europe) with 57 billion EUR in 2014, and 3rd in arrivals with 65 million overnight visitors (UNWTO 2015). Tourism represents more than 10% of GDP, creates 11% of employment and is seen as a strategic axis for recuperation from the economic crisis since 2008 (MINETUR 2013). The Spanish Tourism Plan Horizon 2020 (Plan del Turismo Español Horizonte 2020) takes note that climate change will have impacts on the sector and that actions must be taken to mitigate and adapt to its effects. In order to focus on the sustainability of the touristic model, the Plan foresees the need for:

- indicators to measure sustainability and social and environmental impacts and benefits;
- indicator systems concerning the relationship between climate change and tourism for resources sensitive to climate change (water, air and microclimate);
- methods of analysis and costs of adaptation measures;
- the development of models to evaluate the carrying capacity of the destination using environmental, social and economic parameters.

The 2020 Andalusian Integrated Strategy for Sustainable Tourism (Estrategia Integral de Fomento del Turismo de Interior Sostenible de Andalucía Horizonte 2020) provides the strategic framework for medium term planning for the touristic sector. The 2020 General Plan for Sustainable Tourism (Plan General de Turismo Sostenible de Andalucía Horizonte 2020) provides a greater level of detail than the national level strategy. Its aim at defining strategies and policies to implement a sustainable, equitable and competitive model of tourism that makes the best use of the available resources, respectful of local values and spaces and that guarantees territorial cohesion. There is some awareness of the need to combat the natural (and climate change exacerbated) seasonality of the tourism sector (i.e. the impact of temperature on numbers of visitors) as for example in the 2014 strategy document for the management of tourist seasonality (Estrategias para la gestión de la estacionalidad turística 2014–2020). However, Andalusia does not have a dedicated approach to adapting the sector to climate change. Furthermore, key policies such as the 2020 Andalusian Integrated Strategy for Sustainable Tourism or the 2020 General Plan for Sustainable Tourism provide little mention of climate change and no indication that adapting to climate change is a priority action.

4.3.5 Natural Systems

In Spain the 2008 National Desertification Action Programme (Programa de acción nacional de lucha contra la desertificación, PAND) (in implementation of the

United Nations Convention to Combat Desertification) focuses on the sustainable development of the national territory and in particular the prevention of soil degradation and restoration of land that has been desertified, identifying key factors in its occurrence and practical measures to combat its spread as well as mitigate other effects of drought. The PAND involves an assessment of key factors (aridity, drought, erosion, forest fires, sustainable use of water resources and socio-economic factors) contributing to desertification in Spain. One interesting aspect of this assessment is that it includes an analysis of the impacts—negative and positive— that different sectoral policies and actions of non-governmental actors have on desertification. The PAND focuses on several areas of action such as the definition of measures, coordination of policies and lines of action to fight desertification. For example, it ensures that activities under the 2008 National Irrigation Plan (Plan Nacional de Regadíos)—which focuses on the development and modernisation of irrigated areas—include environmental criteria for the land and water use management to avoid their degradation, the recuperation of aquifers and the reduction of desertification processes. The PAND also includes the definition of areas at risk of desertification and the creation of a desertification Risk Map in Spain. The Risk Map is based on a range of factors and processes including the percentage of surface areas subjected to fires and the overexploitation of aquifers.

In the case of policies governing the use and management of forests and forest fires, the State establishes basic directives at a national level including the 1972 Regulation on Forest Fires (Reglamento de Incendios Forestales) and the 2003 Mountain Law (Ley de Montes). Key policies at national level include the Spanish Forest Strategy, the Spanish Forest Plan and the State Plan for Civil Protection against forest fires. On the basis of these documents the CCAAs must develop their own competencies in areas for which they are responsible i.e. the prevention, detection and extinction of fires.

Andalusia is a frontrunner when it comes to forest management and—along with Extremadura—is one of only two CCAA to have developed a legally binding framework. The pioneering Andalusian Forest Plan (Plan Forestal Andaluz) was approved by the autonomous government in 1989, being the first such effort of its type in Europe. It began with a detailed study of the forest area which identified ten objectives for forest planning. These included *inter alia*, the protection of ecosystems and restoration of those which have been degraded; the protection of forests against fires, plagues and diseases; designation of land use with a view to maintaining its biological potential and productive capacity; rational use of renewable natural resources; improvements to industrialization and commercialisation of forest products; making social, recreational and cultural use compatible with conservation; creation of socioeconomic conditions to avoid the uprooting of rural communities; and diversifying the rural landscape through conservation and recuperation of forest areas within agricultural zones. It is thus a policy which has far-reaching implications, particularly for rural communities. It has a long-term perspective of 60 years, being carried out in decadal phases and revisions in relation to its implementation every 5 years as well as annual progress reports. Forestry planning in the Autonomous Community of Andalusia is integrated into the

Environmental Plan (Plan de Medio Ambiente, PMA) which has the broader remit of promoting sustainable development in the region.

The Forest Plan is complemented in particular by the Plan for Combating Forest Fires in the Autonomous Community of Andalusia (Plan de Lucha contra Incendios Forestales de la Comunidad Autónoma Andaluza, Plan INFOCA)—This action plan which serves the Mediterranean forest area establishes Andalusia as a frontrunner as compared to other CCAAs. Andalusia is one of two CCAA to have passed laws on forests and forest fires: Law 5/1999 on the Prevention and Fight against Forest Fires (Ley de Prevención y Lucha contra los Incendios Forestales), and its Regulation (Decree 247/2001) describe the rights and obligations of the Andalusian administration towards its citizens with respect to regulation of fire use, preventive silviculture and social participation (town councils, farm and estate owners, voluntary groups etc.). It is understood by the Law 5/1999 that a greater level of public participation contributes to an improved situation for forest fires and as such includes various instruments at a local level which promote civil protection for example through the Local Emergency Plans for Forest Fires, Self-protection Plans, Groups for Forest Defence, Groups for Rapid Assistance and collaboration agreements with social organisations. The Regulation also establishes Fire Prevention Plans at a local level with the aim of providing a further guarantee of protection for the Andalusian natural heritage.

4.4   Moroccan Policy Framework

In a review of the Moroccan policy framework carried out by Gerstetter et al. 2011 and indeed since, no policies were found to directly address human security or its connection to climate change and water resources. However, policies addressing climate change adaptation, management of water resources and agriculture can have an impact on human security and related issues such as vulnerability. This section thus provides a broad overview of policies and programmes in Morocco that address water related impacts of climate change and adaptation highlighting their consideration or impact on human security and vulnerability where possible.

4.4.1   Adaptation to Climate Change

Morocco has ratified the United Nations Framework Convention on Climate Change, the Kyoto Protocol and is a signatory of the Paris Agreement. The National Plan for the Fight against Climate Change (Plan national de lutte contre le réchauffement climatique, PNRC), approved in 2009, seeks to reinforce government action to deal with climate change. Coordinated by the Ministry of Environment it brings together all sectors affected by climate change. Morocco has not developed a national adaptation strategy or overriding policy, although this is foreseen (Government of Morocco 2015). The Kingdom is involved in a number of project-based actions to adapt to climate change, many of which fall under Morocco's participation in the global 'Community-Based Adaptation program' financed by the Global Environment Facility's Trust Fund's Strategic Priority on Adaptation, the focus of which remains on capacity building, knowledge communication and research although some have the potential to concretely increase adaptive capacity (Zubrycki et al. 2011). Natural water Reserves such as oases are being affected by lack of water and new programmes are being developed to improve the resilience of the natural habitat for the benefit of local users. This is primarily carried out through the 'Adaptation au changement climatique au Maroc': Programme for resilient Oases resilience (Programme pour des Oasis résilientes, PACC-Oasis) programme sponsored by UNDP as well as through the Territorial Plans to combat Global Warming (Plans Territoriaux contre le Réchauffement Climatique, PTRC). Morocco has also created a High Council for Water and Climate (Conseil Supérieur de l'Eau et du Climat, CSEC) which is responsible for devising general frameworks for national water and climate policy and to examine the coherence of national strategies in relation to knowledge of climate impacts on water resources and management plans. "Reduction of vulnerability to water-related risks and adaptation to climate change" is also included as one of six action plans within the 2009 Water Strategy (see Sect. 4.4.2). Despite the emergence of such efforts to adapt to climate change, Morocco lacks detailed strategic planning of adaptation actions at the national level (Zubrycki et al. 2011) and is still confronted with finding ways to equitably manage the natural spatial and temporal challenges of water supply as described in Chaps. 2 and 3 (see also Box 4.3).

4.4.2 Water Resources

The management of water resources in Morocco is a complex tapestry of interlacing government policies, local customs and traditions (Box 4.3).

Box 4.3. Water rights and management in Morocco
Water rights in Morocco are derived from a variety of sources, including customs, modern laws, and Shari'a rulings. There are rights associated with ASAP (Privileged Farmers Associations/Associations Syndicales Agricoles Privilégiées) rights conceded to the autonomous water distribution state-controlled companies, ancient water rights recognised and registered as well as recognised but not registered (Ouassou et al. 2005). Despite the diversity in water rights, Ouassou et al. (2005) have characterised the current water management context in Morocco as one where "the Administration is still largely in charge of the formulation of policies (…) and the legislative

control is not very well developed", an observation consistent with stake-holder perceptions detailed in Gerstetter et al. (2011).

The centralised control of water resources has long been a feature of Moroccan water policy. In the 1970s, the late King Hassan II launched an ambitious dam policy with the objective of irrigating 1 million hectares by 2020. While the implementation of this policy has undoubtedly led to a greater security of supply, significant disparities exist between basins, and predictions for the future suggest increasing water scarcity as the population grows (OECD 2007). This "aggressive dam building programme" has also been criticised as benefiting only the modern sectors; government-financed irrigation geared to exports benefits a small land-owning elite while tradi-tional agriculture (over 90% of farmers) remains essentially untouched by government policies (Howe 2005)—although this imbalance may now in part be redressed by the Plan Maroc Vert (see Sect. 4.4.4).

In the development of irrigation projects, only recently, through the framework of the Water Users' Associations' Law in 1990 (Association des Usagers de l'Eau Agricole, AUEA), have users been granted a voice in the decision making process (Oassou et al. 2005). These AUEAs are tasked with the responsibility of managing the prioritization and rotation of water access and irrigation as well as the maintenance of equipment. However, according to Bekkari et al. (2008), the creation of AUEAs is non-negotiable, indeed imposed by the State. El Alaoui (2004) (cited in Bekkari et al. 2008) notes that the participation of the AUEAs varies dramatically and is often insuffi-cient and should rather be the result of a real demand from irrigating farmers rather than as the result of a State-led process.

In 1995, Morocco passed its now key legal framework for water management: the Water Law (Loi sur l'Eau) (see Box 4.4).

Box 4.4. Main features of the 1995 Water Law

- An extension of the public ownership of water and the imposition of a time limit of 5 years to any claim on private water rights.
- The introduction of River Basin Agencies (Agences des Bassins Hydraulique) as the main entity in charge of integrated water management at the river basin level.
- The official recognition of State-level planning of availability and allo-cation as the main instrument of decision about public infrastructure, water allocation and water transfer. The Water Basin Master Plan is to be prepared by the river basin agency and submitted to the High Council for Water and Climate (Conseil Supérieur de l'Eau et du Climat) for formal adoption. It then becomes the master plan for water management (includes goals in terms of quality).

- The introduction of new taxes (River basin charge) based on water abstraction, and pollution taxes based on the contribution to the stream pollution. These taxes will, inter alia, cover subsidies to reduce pollution.
- The introduction of new instruments to deal with pollution and drought— fees for polluters, subsidies for investment to reduce pollution and exceptional power to the administration for dealing with drought.
- The formal introduction of the National Hydrological Plan, to be presented to the SCWC, to solve allocation conflicts and make recommendations.

The Water Law provides the context for the development of specific policies and programmes and was, at the time of conception, a particularly innovative framework pre-emptive of many of the measures that would later feature in the EUs Water Framework Directive. The Water Law recognises water as a public good and includes a cost recovery mechanism (via abstraction charges and a water pollution tax based on the polluter pays principle) and the concept of river basins as the functioning unit for water management. The law also reinforces water quality protection by defining environmental mandates and reinforcing sanctions and penalties as well as introducing a 'water police' charged with enforcement powers. Nevertheless, an assessment made by Morocco's Economic, Social and Environmental Council in 2014 found the implementation texts—providing for the Water Law to be either non-existent or completely lacking, with no indication of adaptation to emerging challenges (CESE, Conseil Economique, Social et Environnemental 2014). What is more, numerous articles of the Water Law have failed to be applied, particularly in relation to the overexploitation of water resources and polluter pays principle (CESE 2014; UNECE 2014).

Existing water supply conditions in Morocco, coupled with a growing understanding of climate change impacts on water resources (as outlined in Chaps. 2 and 3) have led the Moroccan government to explore additional policies for ensuring water security. Since the creation of the framework set out by the Water Law, numerous plans, programmes and policies have been conceived with the aim of developing and improving the efficiency of hydraulic infrastructures, access to water resources and sanitation and the integration of water resource management (FAO 2015). However, the sheer number of different initiatives makes the potential for streamlined and coherent coordination questionable. Indeed the Moroccan Economic, Social and Environmental Council found that in practice, the sub-Ministry for Water and the High Council for Water and Climate did not act as coordination bodies and that the Interministerial Commission for Water was inactive (CESE 2014).

A National Water Strategy (Stratégie Nationale de l'Eau) was devised in 2009, and lays out plans to enable Morocco to meet water demands until 2030, notably by mobilising additional investment estimated at 7.5 billion EUR over the 2009–2030 period. The new strategy is characterised by aims to save water and increase the value of that produced. It contains six action plans, one of which addresses the need for demand side management. The expected impacts of this strategy include more sustainable use of water resources through a change in consumer habits and improvement in irrigation techniques, as well as development of high-value added cultivation methods. Importantly for long-term water security, the Strategy also aims to protect river basins and sensitive water resources such as overexploited groundwater Reserves, oases and wetlands and to preserve water quality. One new tool is the development of groundwater contracts for the most important over exploited aquifers between the AUEAs and the government in order to protect the resource and its uses. The first contract was signed for the Souss-Massa in 2011 and another dozen are foreseen (FAO 2015).

The National Water Plan (Plan national de l'Eau, PNE) was originally conceived in 1995 within the context of the Water Law. The PNE established several new public agencies dedicated to securing water supply including the National Office of Potable Water (Office National de l' Eau Potable, ONEP). The PNE encourages the management and development of the supply by increasing the use of non-conventional water resources, including desalinization of 400 million m^3 per year, the reuse of treated wastewater at a rate of 300 million m^3 per year and efficient uses of rainfall water harvesting (Choukr-Allah 2011). However, these measures are currently only implemented on a small scale (FAO 2015). This reuse of water can help meet the high water demands of the irrigation of touristic areas such as golf courses and green spaces. Furthermore it could help to focus attention on a more general need for wastewater treatment: at the national level, 90% of wastewater is currently released untreated into the environment (MEMEE 2010). Consequently there is a separate problem of water quality, which is affected by untreated wastewater entering the cycle. Thus, policies that aim to increase levels of wastewater treatment for alternative uses would also contribute to an improvement in water quality more generally.

Despite this potential, both the Strategy and the PNE continue to rely on the use of large scale water transfers and construction of further dams in order to ensure that water needs are permanently met, and to manage the water contribution of wet years for preparing for drier years. As noted previously, Morocco's water supply is geographically imbalanced, meaning that the government has recourse to mobilise water resources across the country. This is currently executed via 13 inter-basin transfers as well as a network of wells, catchments and traditional systems for water capture and extraction (FAO 2015). However the National Water Strategy and the PNE foresee the construction of a further 59 large dams and around a thousand smaller dams by the horizon of 2030 as well as inter-basin transfers from north to south of 800 million m^3 per year (Simonet 2011). Although a legitimate way to address the spatial and temporal availability of water in Morocco, this type of approach could lead to a type of 'maladaptation' with the country becoming locked into a costly 'hydraulic trajectory' focused on medium and long-term structural measures that cannot be easily adapted with changing climatic conditions. Furthermore, the high levels of siltation from such rapid construction (estimated to be around 75 million m^3 per year (FAO 2015)) means that the total water available is compromised.

Desalination is currently limited to Saharan regions in the South of Morocco (FAO 2015) although it is planned that this capacity should be increased from around 13 million m^3 in 2011 (CSEC 2014) to more than 100 million m^3 by the year 2020 and 400 million m^3 by 2030. There are many other tools that are to hand to address the balance of water supply and demand such as reduction of leaks, progressive pricing, systematic metering and public awareness campaigns which in the case of Rabat-Casablanca have led to a slowing in water demand during the past fifteen years despite high urban growth (El Kharraz et al. 2012).

4.4.3 Floods and Drought

The Department of Water developed the National Plan for Flood Protection (Plan National de Protection Contre les Inondations, PNPCI) in 2002. The PNPCI identified 390 priority locations for which prevention measures are to be realized before 2020. Under this plan Morocco developed a forecasting and flood warning system. The country intends to develop a geographic information system with data on natural and technological hazards across the country, called "GIS-Risk". To realize a "whole-of-society" approach to managing the risk of disaster, Morocco is taking steps towards the creation of a National Platform for Disaster Risk Reduction to support policies for building resilience to disasters.[6] In most river basins, flood risks are currently communicated and coordinated with the Ministry of the Interior and river basin agencies work with local authorities to alert the population on time. Special measures of prevention have been taken in the Ourika basin as this region is a particularly high-risk area and a fully integrated approach has been piloted in this basin. In Morocco, the private market for insurance is not well developed and it is the public insurance that shoulders the main burden of covering natural risks. Morocco is working on a law that addresses natural disasters and is involved in pilot projects for managing flood and drought risk through the impetus of the EU Floods Directive. Morocco has also developed a pilot approach through its Risk Prevention Plans, which it plans to roll out in other vulnerable zones. However plans for adaptation to long-term climate change impacts are not at a very advanced stage (Simonet 2011). Nevertheless, improvement of observation, monitoring and alert tools is coming, with numerous examples of projects, with notable assistance from the World Meteorological Organisation (Simonet 2011).

In terms of drought, Morocco has developed a national programme of action to combat desertification (PAN/LCD) (see Sect. 4.4.6). However, there is little to be

---

[6]The Hyogo Framework for Action 2005-2015: Building the Resilience of Nations and Communities to Disasters—the world's blueprint for creating resilient communities—encourages the establishment of National Platforms for Disaster Risk Reduction. These are multi-stakeholder organizations aimed at improving national coordination in disaster risk management and reduction.

said of a national framework for tackling drought. In theory, the Water Law provides for action to address drought, and Morocco has developed forecasting and prediction systems which have been highlighted as an example of best practice in the region (Troccoli 2008). However, in practice there is a lack of drought early warning systems, planning (as opposed to emergency response) and adapted management in place to ensure water security in times of drought and water scarcity (CESE 2014). Indeed, Morocco was not sufficiently prepared for the extended drought that affected the country in 2015/16 and was forced to launch a new Emergency Drought Plan in January 2016. From the outset, this Emergency Plan is not expected to be able to fully mitigate the effects of this drought episode on the national economy (Jeune Afrique 2016).

4.4.4 Agriculture and Rural Populations

Since the country's independence in 1956, many sectoral programmes, development and reforms have taken place in order to provide food security and contribute to economic growth. In the context of climate change, Morocco's challenge will be to continue to produce food despite increasing aridity and drought conditions. The Plan Maroc Vert (MAPMADA 2008) is the policy that encompasses Morocco's vision for the future of agriculture. The PMV aims to increase the agricultural sector's competitiveness through modernization and integration into the world market whilst at the same time supporting human development objectives. The PMV is based on two pillars (1) the development of a modern agricultural sector based on private sector investment in high productivity/high value added sub-sectors, and (2) the modernisation of production with a social impact targeting small farmers through public investment in social initiatives and *aggregation* (creation of cooperatives) to build solidarity and combat rural poverty. The PMV has developed a list of crop species and their water needs and recommendations for use. The Plan also includes incentives to encourage certain types of crops (water rationalisation). Several programs were designed to support the PMV's implementation. The National Irrigation Water Saving Programme (Programme National d'Économie d'Eau en Irrigation, PNEEI), formulated in 2009, is considered the primary tool in implementing the PMV. With a budget of 3.93 billion EUR, it aims to convert 550,000 ha to drip irrigation by 2020 (Le Matin 2011).

Climate change threatens to jeopardize the gains expected under the PMV. The government of Morocco has attempted to respond to this threat by retroactively 'climate proofing' the PMV through the project "Integrating Climate Change in the Implementation of the Plan Maroc Vert". The latter will finance adaptation measures for small farmers in five regions in Morocco. Ten pilot projects, targeting about 2500 small farmers, will be launched through which farmers will benefit from training and awareness activities. The project, funded by the Special Climate Change Fund (SCCF) and supervised by the Global Environmental Facility (GEF) Secretariat, will also provide training and awareness activities at an institutional level for over 200 staff of institutions involved in the implementation of the PMV (World Bank 2011).

In addition to the PMV, and as mentioned above, Morocco has public insurance, as well as a number of private insurance schemes which are available to assist farmers in times of hardship. However, the take up of these schemes is still rather low. It has been suggested that this is due to low revenue generation by farmers, which in turn does not allow them to join an insurance plan or invest in mutual insurance. In this way, the PMV may help to increase farmers' income, which could in turn develop insurance uptake in rural areas (Oxford Business Group 2009).

4.4.5   Tourism

With 10 million visitors per year and direct employment of half a million Moroccans, tourism is the second most important employment sector after agriculture, accounting for 7.8% of GDP (Government of Morocco 2013). Building on the previous ten-year plan "Vision 2010", the Moroccan government now aims to double the sector by doubling capacity of lodgings and international tourists and tripling domestic tourism with its "Vision 2020 Strategy" (ibid.). The Strategy places higher importance on sustainable tourism by establishing regional monitoring mechanisms and instruments to evaluate sustainability indicators for the industry such as compliance with tourist density thresholds (to avoid ecosystem degradation and adverse effects on local communities), water consumption and the condition of natural cultural attractions. Tourist density levels are being determined for each destination, according to an environmental audit assessing the fragility of the various ecosystems, the resources available and the environmental infrastructure. Furthermore, the government aims to promote responsible tourism through the use of eco-labels and awards that reward those who comply with new industry standards or who show particularly responsible behaviour. The Vision 2020 aims to expand the network of rural and eco-tourism destinations. The intent is also that these types of nature tourism will increase income generating activities for local communities, reducing the effects of poverty in rural areas and creating new employment opportunities for young people (Roudies 2013). Nevertheless, there is a strong need for this strategy to be appropriately monitored and implemented to ensure that it achieves its aims of growth without compromising people or environment, particularly in the context of reduced water resources as a result of climate change.

4.4.6   Natural Systems

The National Action Plan to Combat Desertification (Programme d'Action National de Lutte Contre la Désertification, PAN/LCD) was passed in 2001 in response to the UN Convention to Combat Desertification (UNCCD). A High Commission (Haut Commissariat aux Eaux et Forêts et à la Lutte Contre la Désertification, HCEFLCD) was established to manage this programme, establishing a decadal programme made up of territorial projects. The (2005–2014) programme focuses on three strategic objectives (i) combating desertification (ii) the preservation and development of forest resources and (iii) human development in forests and surrounding areas. In order to attain these objectives, a programme of action was devised. Areas of action include halting erosion, reforestation and regeneration of sylvo-pastoral systems, the contribution of forest spaces to improve living conditions for surrounding populations, conservation of biodiversity and sites of biological and ecological interest and improving the contribution of local forestry products to cover the needs of local industry and handicrafts. The next decadal programme (2015–2024) includes a strong focus on reforestation, particularly through the mobilisation of local populations into cooperatives and associations to help defend forest areas. Management committees and participatory management plans are also foreseen for protected areas to conserve biodiversity and develop eco-tourism.

From 1942 to 2006, Morocco created nine national parks to protect its fragile and insufficiently protected natural systems (HCEFLCD 2015). The Directive on Protected Areas (Plan Directeur des Aires Protégées) was established in 1996 with a focus on the conservation and rational use of natural resources and fight against desertification. Furthermore, they promote the development of economic activities considered compatible with their conservation objective such as tourism or eco-tourism. In addition to its national parks, Morocco has three Biosphere Reserves, which marry biodiversity conservation with sustainable use. One of these, the IBRM is the subject of this study.

4.4.7 Human Development

Human development is connected to human security in a number of ways. It represents an attempt to reduce the vulnerability of populations to shocks by providing citizens with greater opportunities for education (which can increase economic security) and access to food, water and sanitation (which contributes to food and water security and health). Launched in 2005 by King Mohamed VI, the main objective of Morocco's National Human Development Initiative (INDH) was the reduction of socioeconomic deficits by satisfying the fundamental needs of the poorer parts of the population via the following four axes of implementation (Abdesslam 2011): (1) The fight against rural poverty; (2) the fight against urban social exclusion; (3) the fight against vulnerability; and (4) intersectoral programmes. For the period 2006–2010, more than 200,000 actions and development projects were initiated for a total budget exceeding 13 billion MDH (1.2 billion EUR) (ibid.). The NIHD gives significant weight to the issue of water; Morocco has implemented a series of strategies to increase access to water in urban and rural areas. For example, policies implemented by the operators in Casablanca, Rabat, Tangier, and Tetuan allowed low-income families in peri-urban settlements to have access to a water connection. The connection ratio was expected to reach 100% by 2015 (OECD 2007). Although access to improved water sources was estimated to be at 99% coverage for urban settlements in 2015, in rural areas this was still limited to 65%, with only 23% of the rural population having water piped onto the premises (WHO/UNICEF 2015).

In 2010, the National Strategy for Sustainable Development [Stratégie Nationale de Développement Durable (SNDD)] was conceived as a new legislative framework to pursue environmental standards in policies across the board. In this way, it will assess key sectoral strategies for their sustainability, such as:

- the PMV and its objective of significantly increasing agricultural yields and the maintenance of its social impact for rural populations through added value projects, while playing a role to protect natural resources, notably through water saving methods in agriculture, the reduction of soil pollution and the promotion of clean energy in agriculture.
- the Tourism Vision 2010 and 2020 and their environmental objectives, in particular in the development of tourist complexes, the deployment of the Moroccan responsible tourism charter and the promotion of touristic eco-labels.
- INDH and its objectives of poverty reduction and improvement of quality of life, limiting the pressure on natural resources.

4.5 Policy Frameworks in the IBRM

The key policy frameworks relevant to the IBRM are those described above that have been conceived at supranational, national or regional level. A more specific legal framework or common policy for the IBRM itself does not exist. As noted above, however, there is an Action Plan (2011–2015) and a Memorandum of Understanding on the Coordination of the IBRM signed by the Spanish and Moroccan governments. While there is a commitment to joint management from both sides, financial support from Spain is often a prerequisite to action on the Moroccan side (personal communication, stakeholder workshop). A training session for managers and technicians of the Reserve was offered in November 2015 by the Spanish government to explore the applicability of the system of indicators in the Spanish Network of Biosphere Reserves to Moroccan reserves with "significant advances" made to this end (Spanish National Report 2016). The session also discussed the establishment of participatory bodies in biosphere reserves in Morocco and the planning of the ten-year evaluation of the IBRM, scheduled for 2017 (ibid.).

A new generation of projects with co-financing have also emerged between Morocco and the Spanish agency for International Development [Agencia Española de Cooperación Internacional para el Desarrollo, (AECID)]. Local NGOs in

Chefchaouen, the 'Association de Développement Local de Chefchaouen' and the 'Association Tallasemtane pour l'Environnement et le Développement' have several local projects on the effects of climate change, rural tourism, flooding, improvements for the socio-economic situation of the local community.

4.6   Stakeholder Insights

The perceptions of stakeholders—both those directly engaged in the creation and implementation of policy, as well as those affected by policies—are crucial to the assessment of policies (Bryson 2003; Brugha and Varvasovszky 2000). The following section details insights gained through two stakeholder focus group meetings held in February 2012 (see also Sect. 3.4). The first on the Spanish side of the IBRM in Malaga on 13th February 2012 and the second on the Moroccan side in Tetuan on 15th February 2012. The main points raised in discussions on Scenarios A and D are summarised in four tables in Annex B. This section takes a narrative approach to describe the broad findings of these discussions, after which their implications in relation to the policy frameworks outlined above are discussed. Although the conclusions from the stakeholder workshops cannot be said to be representative of all interventions made by each participant, care has been taken to represent the majority viewpoint and to focus on issues that were raised by more than one stakeholder.

4.6.1   Spanish Side of the IBRM

On the Spanish side of the Reserve, the context of climate change was considered to be an opportunity to take positive planning action to improve the future situation of the IBRM. Here, discussions also covered a wide range of measures for effective water management in the face of future climate change scenarios. These included: water treatment and use of alternative sources of water (waste-water reuse); improvement of infrastructures for increased storage capacity and water saving measures to increase the availability of the resource; artificial recharge of groundwater; improvement of the quality of the resource to ameliorate the state of the ecosystems; use of appropriate tariffs for water which could be reinvested in the IBRM and demand-side management. Ways in which management could be improved were suggested, in particular the need for greater flexibility in water resources allocation and prioritization of sectors according to needs. Similarly to the Moroccan side of the IBRM, implementation was a key issue. The application and monitoring of the implementation of the international, national and regional legislative framework for water regulations in force (hydrological plans, Water Framework Directive, Water Law etc.) was seen to be key and in need of strengthening. One specific suggestion was the creation of a public watchdog that could be tasked with carrying out the monitoring of the use of water resources as well as the creation of an entity solely tasked with the cross-cutting issue of water management with a view to better coordination between bodies and departments in the administration.

Particular attention was paid by the Spanish focus group to the role of civil society. It was noted that there was no tradition of public participation and that there was a lack of interest due to the fact that final decisions are made at a political level. The Water Framework Directive was identified as promoting consultations rather than participatory processes. In this way the need for improved capacity building for citizens was highlighted as a way of increasing co-responsibility in water management. Furthermore, it was suggested that citizen education would assist in promoting the sustainable use of resources, conservation and the perception and use of alternative sources of water such as reused waters. The promotion of further Research, Development and Innovation to supplement knowledge gaps was highlighted as an area for action, particularly at European level.

Other measures discussed to address climate impacts in the IBRM included the actions in emergency response, tourism and agriculture. For emergency response, a need was seen for improved preempting of the situation, focusing on emergency rather than disaster management. Suggestions for improving this policy area included Research, Development and Innovation in early warning systems and the design of indicators which allow extreme situations to be preempted, and the creation of municipal emergency plans (at town council level). Similarly to the case made for water taxes and tariffs, it was suggested that tourist taxes be taken and directly reinvested in the IBRM. Furthermore the need for sustainable tourism was emphasised in particular in relation to water use; the water perspective was seen to be lacking from current tourist policies and approaches. Finally, for agriculture, policies to incentivise particular types of crops that are better adapted to the climate were seen to be necessary. The need for food security in order to provide food for tourism was also mentioned in this context.

4.6.2 Moroccan Side of the IBRM

The stakeholder proposals for water management in the Moroccan focus group were multi-faceted, with many different measures put forward for addressing the impacts of climate change. These included: desalination, dam building, wastewater and residual water reuse, water saving measures such as demand reduction, improvement of water infrastructures, development of new technologies as well as greater efficiency in irrigation. In this way, participants' priorities largely mirrored the aims set out in the existing Moroccan policy framework. Methods suggested for carrying out these measures included taxation on water consumption, promotion and increased economic support (i.e. through subsidies) of more efficient approaches and technologies such as drip-irrigation, public awareness campaigns, new legislation to account for climate change, and above all, the more effective implementation of existing policies. This latter point was raised on several occasions. Despite Morocco signing 'all' the international conventions and having a relatively expansive policy and legal framework, stakeholders found that application of these policies remained poor. In relation to impact assessments, for example, there was seen to be a gap between conception of the assessment and its execution, with the indication that impact assessments were not being carried out for key policies. Obstacles to ensuring current and future water security identified by participants included: exhaustion of water resources for the intensification of export crops (a focus of the PMV); the siltation of dams (a consequence of the government's rapid and large-scale dam-building policy) and the high costs and energy consumption of desalination (plans to increase substantially over the coming years). In addition, more resources were called for to improve integration between sectors and research fields and to increase levels of openness and collaboration. In the specific case of the PMV, although it was seen as being of potential interest by farmers in the IBRM, the policy is designed for plots of land larger than 5 ha and is less relevant due to the fact that most areas of the IBRM are owned by smallholders with around 2 ha.

Although the need for further research on water resources (knowledge of available surface and groundwater, water efficiency measures) was mentioned, it was also made clear that a great deal of traditional knowledge and practices could be mobilised for the purposes of rediscovering climate resilient crops and varieties and other solutions to water management. Indeed, it was pointed out that greater awareness and involvement of civil society in water management and their engagement with the research community and the public administration would assist more generally in managing current and future climate impacts on water (although real levels of support for public participation were held in doubt by the focus groups). It was also suggested that economic profits from water should be shared and invested in the protection and management of the resource for the local public.

Parallel to water related measures, protection of the natural environment, conservation of biodiversity and reforestation in particular were seen to be of importance. Furthermore, the need was highlighted for reforestation and agriculture to be carried out with species and varieties that consume less water. This aspect was recognised as an area of opportunity, given that there is local knowledge on the types of varieties that might be better adapted to the expected changes in climate.

One interesting, and unexpected aspect of the debate surrounded the growth of cannabis crops, largely for export to Europe. This illegal 'cash crop' (currently) grows well in the IBRM and provides higher incomes than many other crops. However, the effect of its cultivation (usually through monocropping) has both negative social and environmental impacts. Suggestions for addressing the issue included its legalisation for medical use, the control of external demand and increasing the number of prosecutions. According to participants in the focus groups, further economic diversification could take place through promotion of local or cooperative products (herbs, jams, honey and handicrafts) and activities (rural and sustainable tourism e.g. bed and breakfast, cycling tours) as an element of the touristic services on offer.

4.6.3 The IBRM

Stakeholders made little reference to the IBRM policy framework per se (e.g. the Action Plan), and tended to focus on their national policy context. On both the Spanish and Moroccan sides, stakeholders considered that current policy frameworks were generally appropriate but that there was a lack of appropriate implementation mechanisms and a need for greater monitoring and control. On both sides, the need for cooperation between research, civil society and administration was highlighted. On the Spanish side, this was above all to facilitate the transfer of know-how from universities and technical experts to administrations, implementing agencies and populations in rural areas. On the Moroccan side, emphasis was placed on the local knowledge already held within the IBRM communities and the need to increase the value and use of this type of know-how in a more bottom-up process. Both sides highlighted the need for public awareness-raising, particularly in relation to wastewater re-use, the protection of the environment and the need for demand-side management. Civil engagement was seen to be a positive goal, however, in both Spain and Morocco past experiences have led to low-expectations in this actually being implemented.

4.7 Conclusions

The most obvious and contrasting feature of the Spanish policy framework as compared to Morocco is its membership of the European Union. The EU determines the framework for the development of water related policies and climate change adaptation management in Spain and Andalusia. In the case of Morocco, the role of a supranational framework is played to a certain extent by the International Conventions it has signed. Nevertheless, in both Morocco and Spain, these supranational frameworks primarily relate to the initiation of the process of law and policy-making. Once these frameworks have been incorporated into Spanish or Moroccan law, they have the same effect as those of any other part of the national policy framework. One key distinguishing factor however, is that Spain is required to monitor and report on progress in the transposition and implementation of EU laws and is subject to penalties should it fail to meet requirements. As for Morocco, it is not subjected to the same consequences for failure to carry out pledges made in the context of international agreements, treaties and conventions.

Both Spain and Morocco have made considerable efforts in the last decades to develop their policy frameworks in response to water-related challenges, both from the perspective of water policy as well as through other sectors such as agriculture, forestry or tourism. The Spanish approach to drought management is particularly impressive. Nevertheless, there are also gaps within these frameworks which must be filled. For example, while there is recognition at national level in Spain of the need to adapt the tourism sector, there is insufficient focus on adaptation at the

Autonomous Community level. From the Moroccan side, there are numerous efforts to tackle water resource management. However, these policies are for the most part severely lacking in implementation and there is an absence of strategy in relation to climate adaptation which is currently rather piecemeal in its approach.

Nevertheless, on the whole, policy frameworks in Spain and Morocco do increasingly consider the risks posed by climate change to water and the implications for the vulnerability and security of human populations. Indeed, stakeholders considered that, on paper at least, the present institutions and policies were adequate for addressing the challenges faced by the IBRM. However, in order for these frameworks to be truly effective, there is a strong need to focus on the local implementation of these policies and to simplify existing institutions to reduce overlap and duplication of efforts. Stakeholder suggestions for improving the policy framework in the IBRM (Box 4.5) frequently mentioned the need to strengthen public participation and civic co-responsibility in water management. Indeed, this is a goal that is positive both in the spirit of democracy and transparency as well as its potential to support the process of policy implementation.

Box 4.5. Stakeholder suggestions for improvement of policy framework in IBRM

Morocco	Spain	General
• Integrate local knowledge in development planning (empowerment of civil society) • Implement plans for reforestation and preservation of natural systems • Promote the change to low water consumption irrigation systems e.g. drip-fed • Use alternative water sources in extreme scenarios, such as desalinisation and wastewater reuse • Change to crop species and varieties adapted to drought conditions • Reinvest economic benefits of the water sector in the conservation and protection of natural resources	• Improve the management of natural and forest areas • Improve the existing water infrastructure • Create monitoring networks • Consider alternative water sources • Improve the technical capacity of the public administration	• Fulfilment of current laws and normatives and enforce their accomplishment • Encourage public participation and population co-responsibility in water management • Promote more efficient water technologies and solutions • Improve the link between research, public sector and private sector • Introduce touristic taxes and effective water pricing • Strengthen the IBRM common institutions • Promote ex ante adaptive measures rather than ex post measures (crisis management rather than catastrophe management)

The IBRM currently has a management plan, however this is a loosely formed agreement and there is no associated governing body. Thus management of the Reserve is carried out in a parallel process through the respective policy frameworks in Morocco and Spain. In this way, the Reserve would benefit from a common management framework with strengthened institutional coordination and budget. Such a framework would need to be adapted to both sides taking into account their differing physical and human contexts (see Chap. 2). However, these efforts would strengthen the capacity of the Reserve to address the impacts of global and climate change on the natural systems and human populations and guide the sustainable planning, management and control of water resources at the river basin level.

Chapter 5
Vulnerability Assessment

Abstract Climate change is expected to have severe impacts on the Mediterranean region. However, the extent to which socio-ecological systems are vulnerable to these impacts is defined by the level of preparedness and the capacity to respond and adapt. In this chapter an integrated vulnerability assessment is carried out for the Intercontinental Biosphere Reserve of the Mediterranean (IBRM). The assessment takes into account the exposure of the IBRM to the hazards of climate and global change, the impacts of these changes, and the adaptive capacity of political systems to cope with the potential negative consequences for human security. The IBRM presents a particularly interesting case study for this assessment. On the one hand, the current and future effects of climate change are rather homogenous throughout the Reserve which is ostensibly protected under a single framework. On the other hand, the Reserve extends across two separate nation states (Morocco and Spain) with very different institutional and cultural contexts. In this way, both the manner in which climate change affects socio-ecological systems and the capacity of policy frameworks to address the impacts vary considerably. The assessment finds that there are considerable differences in the type and extent of vulnerability in each national jurisdiction of the IBRM.

Keywords Mediterranean basin · Spain · Morocco · Climate change · Climate adaptation · Adaptive capacity · Human security · Food security · Water security · Floods · Water scarcity · Drought · Tourism · Agriculture · Policy frameworks · Hydro-ecological model · Scenarios · Impact assessment · Vulnerability assessment

5.1 Review of Vulnerability Assessment Approaches and Methods

In 2008, the UNFCCC Secretariat developed a compendium of methods and tools to evaluate vulnerability and provided a chronological development of vulnerability assessment approaches and methodologies. The Secretariat observed that first

© The Author(s) 2017   61
D. Abdul Malak et al., *Adapting to Climate Change*,
SpringerBriefs in Environmental Science, DOI 10.1007/978-3-319-51680-6_5

generation methodologies focused more on impacts than on adaptation, as it was fundamental to quantify the impacts to drive the mitigation agenda and the stabilization of greenhouse gas concentrations. These studies applied climate scenarios based on general circulation models (GCMs) to ecosystem models to assess climate impacts on species, ecosystems or biogeophysical environments. Some studies went a step further by also assessing the impacts on economic sectors (e.g. agriculture, forestry, water resource management). Later, climate scenarios were complemented by the inclusion of socioeconomic scenarios. The second generation of methods linked adaptation with current climate variability and vulnerability in addition to future climate and vulnerability, widening the scope to other environmental, economic and social stressors. Methodologies are evolving rapidly and the most recent approaches include aspects such as stakeholder participation, adaptation policies and measures or adaptive capacity.

There are several approaches to vulnerability assessment, which depend on the framework, assumptions, characterisations and value-sets being applied (Lynch et al. 2008). For example:

- Fussel (2007) characterises five types of vulnerability assessment framework: (1) the *risk-hazard* approach, applied mainly in research and disaster evaluation, assesses the level of risk of a physical system to the exposure of a hazard; (2) the *political economy* approach, frequently used in poverty and development studies, assesses who is most vulnerable and why. Here vulnerability denotes the socioeconomic response capacity of individuals and groups to a variety of stressors; (3) the *pressure-and-release* model starts from the risk-hazard framework, defining risk as the product of hazard and vulnerability, and involves global root causes, regional pressures, and local vulnerable conditions; (4) the *integrated* approaches, as for example the *hazard-of-place* approach, assess vulnerability as an integration of the exposure of a system to natural hazards and its adaptive capacity. These integrated approaches are widely used in the context of global environmental change and climate change; and (5) the *resilience* approach which assesses vulnerability as the propensity of social and ecological systems to suffer harm from exposure to external stresses and shocks, involving exposure to events and stresses, sensitivity to such exposures and resilience owing to adaptive capacity measures to anticipate and reduce future harm.
- Lynch et al. (2008) propose four different approaches: (1) the *hazards, natural hazards or risk approach* assesses vulnerability as the biophysical changes produced in a system; (2) the *vulnerability* approach assesses vulnerability in terms of the values at risk to the hazard; (3) the *adaptive capacity or resiliency approach* focuses on the availability of resources or the ability of a biophysical system to cope with change; and (4) the *policy* approach assesses how vulnerable a policy is to change.

- The UNFCCC Secretariat (UNFCCC Secretariat 2008) categorises depending on how the tools or approaches address key cross-cutting themes or address complete frameworks, classifying: (1) Complete Frameworks and Supporting Toolkits; (2) Cross-Cutting Issues and Multisector Approaches (general tools, decision tools, stakeholders approaches and other multisector tools); and (3) Sector-Specific Tools (agriculture, water, coastal resources, human health sector and terrestrial vegetation sector tools).
- The U.S. Environmental Protection Agency (EPA 2010) classifies the approaches as either *top-down* modeling assessments or *bottom-up* threshold analyses. It describes the *top-down* assessment as model and data driven, being more time and resources consuming. The *bottom-up* approach is a more qualitative analysis based on expert knowledge.
- The San Francisco Bay Conservation and Development Commission (BCDC 2011) distinguishes between four approaches: (1) the *vulnerability-risk-response* approach; (2) the *risk-response* approach; (3) the *exposure-response* approach and (4) the *vulnerability-response* approach.
- Hammill et al. (2013) makes a comparative analysis of vulnerability assessments in Tunisia and Indonesia, observing three different approaches: (1) the *risk-hazard (RH)* approach assesses vulnerability as the expected consequences (losses) of exposing sensitive people and/or property to a particular (external) hazard. This approach is typically associated with 'top-down' or scenario-driven vulnerability assessments (2) the *political economy* approach focuses on the socio-economic processes that lead to differential exposure, impacts, and capacities to deal with impacts. This approach tends to be characterised as 'bottom-up' (3) the *integrated approach* assesses vulnerability as function of the exposure, sensitivity and adaptive capacity of a system to a stressor. This approach is used by the Intergovernmental Panel on Climate Change (IPCC).

Despite the differences between the above mentioned approaches, it is also important to note several similarities: (1) the majority state the need to define the scope of the assessment and to agree upon the definitions and frameworks; (2) many engage stakeholders and those who will be most affected by the impacts of climate change; and (3) the majority encourage the use of climate scenarios to reduce uncertainty in estimating future vulnerability.

Most studies apply the methodology proposed by the IPCC (2007b) assessing vulnerability as a function of the character, magnitude, and rate of climate variation to which a system is exposed, its sensitivity, and its adaptive capacity. However, the 2014 IPCC report recognises that vulnerability has acquired increased complexity as a multidimensional concept, with more attention needed for examining the relationship between vulnerability and the structural conditions of poverty and inequality.

The majority of vulnerability assessments in the Mediterranean basin assess vulnerabilities related to coastal areas (Nicholls and Hoozemans 1996; Snoussi et al. 2008; UNEP-MAP RAC/SPA 2009; Otero et al. 2013; Larroudé et al. 2014) or biodiversity (Gritti et al. 2006; Jordà et al. 2012; Filipe et al. 2013).

Mediterranean vulnerabilities are also included as examples in research projects and reports with a broader focus e.g. Europe (EEA 2005; Schröter et al. 2005; Lindner et al. 2010) or global (IPCC 2007, 2014). A previous example of a more comprehensive vulnerability assessment for the Mediterranean is the CIRCE project (6th Framework Programme 2007–2011), which predicted and quantified the physical impacts of climate change in the Mediterranean, assessing the most influential consequences for the regional population. The results of the project are summarised in the book *Regional Assessment of Climate Change in the Mediterranean* (Beniston 2013).

5.2   Integrated Assessment of Future Ecological and Social Vulnerabilities of the IBRM

This book applies an integrated approach to vulnerability assessment (Fussel 2007) to include both biophysical and socioeconomic factors and is informed by approaches used in the disaster risk community. The assessment takes into account the IBRMs exposure to the hazards of climate and global change, the impacts of these changes, and the capacity to cope with the potential negative consequences. Both the impacts and the adaptive capacity thereby determine the level of vulnerability and potential consequences for human security. Figure 5.1 shows the conceptual framework we have followed for the assessment of vulnerabilities in the IBRM.

Fig. 5.1 Conceptual framework of the vulnerability assessment applied in the IBRM

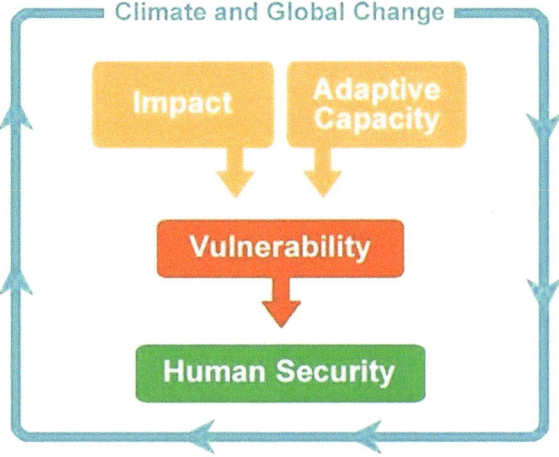

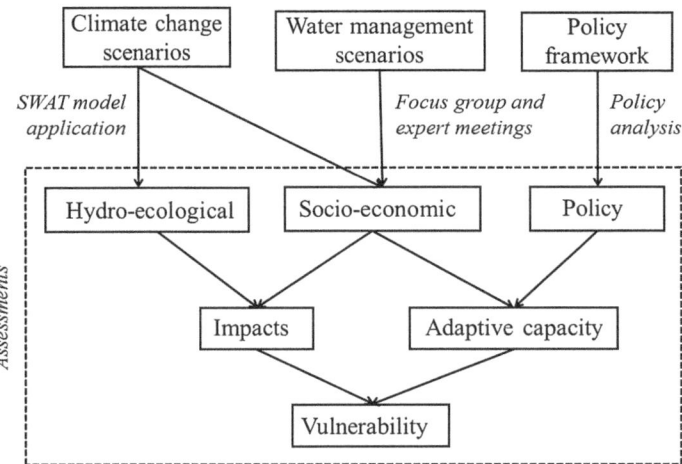

Fig. 5.2 Methodological framework of the vulnerability assessment applied in the IBRM

5.2.1   Approach and Methodology

The assessment takes a multi-disciplinary approach to consider how different dimensions contribute to the vulnerability of the IBRM. In the previous chapters, three assessments have been carried out: hydro-ecological; socio-economic; and policy assessments. The impacts depicted in Fig. 5.1 were defined on the basis of a hydro-ecological assessment and a socio-economic assessment (Chap. 3). The hydro-ecological assessment was conducted on the basis of climate change scenarios and the application of a hydro-ecological model (SWAT). The socio-economic assessment was based on water management scenarios and inputs from focus groups and expert meetings with stakeholders. The adaptive capacity depicted in Fig. 5.1 was assessed on the basis of a policy assessment where policy frameworks were analysed on the basis of desk-based research and stakeholder inputs (Chap. 4). The current chapter integrates the results of these three assessments to carry out an analysis of the overall level of vulnerability on each side (Spanish and Moroccan) of the IBRM. The methodological framework is illustrated in Fig. 5.2.

5.2.2   Vulnerabilities in the Physical Environment

5.2.2.1   Climate

The Mediterranean region is known for its seasonal contrasts and is characterised by a high annual variability in precipitation and frequent drought episodes, especially during the summer months. The population living in dry areas, especially marginalised groups, is likely to be most seriously hit by the shifts in moisture and temperature regimes (Karas 1998; UNFCCC 2007). These increasing climatic pressures, especially the temperature increase, have both positive and negative effects for the IBRM (Plan Bleu 2008; Schröter et al. 2005; Tellería et al. 2016). On the one hand the changes produce a reduction in energy consumption for heating during the cold season; on the other hand they increase the cooling requirements in summer. Nevertheless, the decrease in precipitation and changing rainfall patterns are perceived to be more problematic as they imply less water availability and more flood events respectively (Giorgi 2006). In **Spain**, the National Climate Change Adaptation Plan (PNACC) has identified water as a priority area, pushing for the creation of a framework which attempts to address the cross-cutting nature of global change across different sectors. The National Desertification Action Programme (PAND) includes the definition of areas at risk of desertification and has also led to the creation of a desertification Risk Map for Spain. At the level of implementation, the Spanish governance system—with competences at Autonomous Community (CCAA) level and low participation of civil society—is still considered to be a major challenge for the effective implementation of such measures. In **Morocco**, although some projects to pilot resilience to climate change exist (e.g. for oases), the lack of a detailed and strategic plan for climate change adaptation is considered to be a major hindrance.

5.2.2.2 Natural Disasters—Floods

Expected changes to patterns of seasonal precipitation will increase the risk of flash floods in the IBRM. This will increase in intensity further downstream, impacting coastal regions outside the boundaries of the reserve. Floods, exacerbated by deforestation and inappropriate agricultural practices are having a direct impact on soil erosion (IPCC 2007a); annual soil losses of over 4000 Mg/km^2 have been registered in the Rif province of northern Morocco (Ouassou et al. 2005) and threaten 21% of the soil in the province of Malaga in southern Spain (RME 2008). In the context of implementing the 2007 European Floods Directive, **Spain** has a comprehensive framework and timeline for increasing flood risk protection. The Spanish Automatic Hydrological Information Systems assist with the prediction and management of floods. Spain has also produced risk maps and flood management plans which—as with drought management—are now focused on prevention, protection and preparedness. However, these plans were released in 2016 and as such, there is not yet evidence of their effectiveness in reducing impacts and overall vulnerabilities. Extreme flash floods in the first decade of the 21st century in **Morocco** led to dozens of fatalities as well as substantial damages to personal and commercial property and land (World Bank 2014b). The government has developed a forecasting and flood warning system and intends to develop a GIS Risk system that takes a whole of society approach to managing Disaster Risk. Nevertheless, this is still in the process of development and its effectiveness for mitigating future flood events is not yet clear. Furthermore, the aforementioned autonomous adaptive responses to drought in Morocco may further add to flood risk. The impacts of drought on rural livelihoods increase both seasonal and permanent migration towards coastal areas with economic opportunities e.g. in tourism (ENPI 2007). This results in population increases in these vulnerable areas, and thus an overall increased risk to populations from the impacts of coastal floods.

5.2.2.3 Natural Disasters—Droughts

In the latter part of the twentieth century, Spain and Morocco experienced extensive and recurrent droughts. Many arable agricultural systems (both rain-fed and irrigated) had to cease production due to insufficient water reserves (CRED 2010). The droughts also led to serious losses in pastoral agriculture and in the forestry sector in Spain (Peñuelas et al. 2001), and in Morocco (Martin et al. 2016). Drought episodes are likely to be exacerbated not only by climate change impacts but also by growing water demand resulting from socio-economic factors such as population increase, land use change and the further expansion of water-intensive sectors such as tourism and agriculture (COM 2016; Magnan et al. 2009; Plan Bleu 2008, ENPI 2007; Benoît and Comeau 2005). In **Spain**, 6 million people were affected by the droughts of the mid-1990s, and many migrated from rural areas to urban centres in order to ensure adequate supply of drinking water. However, in the last decades, the country has developed its expertise on drought as well as its capacity to respond. Spain has moved from a crisis-centred approach to drought towards an approach centred on planning and prevention. This has led to the design of a national drought indicator system and the creation of Drought Management Plans (DMPs), which contributed to the alleviation of the most recent drought episode (2004–2008) (Estrela and Vargas 2012). Moreover, the Spanish approach has been highlighted as an example of good practice and has had a strong influence on the formulation of drought policy at a European level. In **Morocco** a range of locally adapted indigenous systems for water storage and regulation exist to counteract general water scarcity (e.g. Kutsch 1982). Rural populations also carry out autonomous adaptation to situations of reduced water availability or drought by temporarily migrating to urban areas. However, the national economy is highly reliant on the agricultural sector; in 1995, drought caused GDP to fall by 7.6% (RMSI 2012). In this context, autonomous adaptation is no longer an adequate response for ensuring food and economic security. Although Morocco has developed systems of forecasting and prediction (e.g. the Direction de la Météorologie Nationale—lauded as a regional success story (Troccoli 2008; El Khatri and El Hairech 2014), it has yet to develop generalised systems for planning and drought early warning. Indeed, the lack of planning and the prevention have led to the need to launch a new Emergency Drought Plan in January 2016 to address the drought episode of 2015. From the outset, it is not expected that this Emergency Plan will be able to fully mitigate the effects of this drought episode on the national economy (Jeune Afrique 2016).

5.2.2.4 Water Resources Vulnerabilities

Official records of inter-annual climatic variability paint a picture of natural water scarcity with reduced water resource availability in the Mediterranean region (IPCC 2007a). Coastal and low altitude areas are subjected to greater pressures than mountainous or humid areas and this situation is expected to worsen in the future. In parallel, water demand has been growing, mainly due to the development of tourism and the intensification of agricultural production, putting more pressure on available water resources (ENPI 2007; Benoît and Comeau 2005). Due to the seasonality of sectors such as tourism, water shortages are expected to be exacerbated at particular times of the year (Méndez 2008). At the policy level, the implementation of water policy requires better coordination between responsible bodies for water use and management in both countries. Besides this, the prioritisation of sectoral water uses (i.e. to give priority to meet water demands from one sector at the cost of other sectors) needs to be better coordinated and adapted to the realities of climate change. The increasing pressures on natural resources on the **Spanish** side of the IBRM are posing a threat to the sustainability of ecosystem services provided to local population, especially in coastal zones. This fact will be accentuated by climate change, given that hydrological models foresee a notable reduction in water stream flows. This reduction will affect water quality, stimulating an increasing need for the development of costly new and more efficient water technologies. Water management is perceived as a top down activity with low levels of public participation. However, adaptation requires awareness in the population to favour demand reduction and question unsustainable infrastructures (e.g. golf courses) and improve the perception of alternative water management practices (e.g. use of wastewater). Water regulations are perceived as strict and rigid and will need to increase in their flexibility in order to adapt the water sector to global change. Due to the level of opposition to the National Hydrological Plan (NHP), large scale water transfers are not feasible in Spain. The AGUA programme implements the Water Framework Directive (WFD) in principle although full cost recovery is still an outstanding question. Spain has a long experience with river basin level management but greater public participation is needed to allow for historical water rights to be reallocated based on new criteria such as equity, efficiency and sustainability. Global change effects on the **Moroccan** side of the IBRM have led to serious impacts on the quality and availability of water resources including the malfunction or interruption of drinking water treatment plants, the limitation of irrigation water supply and the increase of fish deaths and waterborne diseases (Ameziane et al. 2001). Future scenarios foresee a notable reduction of water stream flows. This situation is worsened by inadequate implementation and monitoring of policies, or the creation of policies that are in effect examples of maladaptation e.g. large scale water transfers, rapid dam building leading to siltation, inadequate use of alternative water sources, lack of controls of the discharge of untreated waters, the promotion of luxury tourism and the promotion of high value crops for export that require substantial levels of irrigation. There is also a lack of involvement of relevant stakeholders to determine the management and prioritisation of water use. In this way, it is necessary to fully consider the interplay between different sectoral policies in order to ensure effective adaptation to climate change impacts and the reduction of vulnerabilities.

5.2.2.5 Natural Systems Vulnerabilities

The IBRM and surrounding areas have experienced a process of land degradation caused by: rapid socioeconomic growth; the increase of urban and touristic development (e.g. Costa del Sol); and technological transformation (e.g. the Tanger-Med cargo and passenger port). Besides this, the expected increase in environmental aridity is expected to have impacts on forests and natural ecosystems, causing losses in ecosystem integrity and biodiversity, forest decline and forest fire (Jump et al. 2006; Sarris et al. 2007; Moriondo et al. 2006). The models applied in this study predict a higher potential vulnerability of forests to future climate trends than shrublands as forested areas may see a greater reduction in soil water storage than shrublands. For this reason, a possible expansion of shrubland distribution and range in the region is expected. 70% of the **Spanish side of the IBRM** is covered by natural systems (forests, shrublands and grasslands) (GlobCover v.2.3. 2009, 2010) of which 70% is protected area (Molina and Villa 2008). The foreseen increase in temperature may cause different impacts: (1) the reduction of cork production, damaging the forest sector; (2) the increase of fire risk; (3) the spread of forest diseases; and (4) the intrusion of alien species. Moreover, biodiversity and natural ecosystems will suffer from ecological flow shortages and the eutrophication of water reservoirs. Andalusia has been a pioneer in the development of management policies for forest protection, with the Andalusian Forest Plan (approved in 1989) being the first effort of its kind in Europe. Andalusia has also developed policies to prevent forest fires, in which civic engagement is promoted together with various instruments at local level to promote civil protection. Forests and natural systems occupy around 39% of the **Moroccan side of the IBRM** (GlobCover v.2.3. 2009, 2010), and only 30% of this area is protected (Molina and Villa 2008). Although they occupy a smaller surface area in the Reserve, Moroccan forests appear to be more sensitive to future aridity condition than their Spanish counterparts. On the Moroccan side, there is a greater proportion of shrubland than forests. On the one hand this distribution is in tune with the climatic conditions and requires lower levels of water. On the other hand, shrubland does not make any positive contribution to combat erosion and soil degradation, contributing to the overall vulnerability of the system. Similar to Andalusia, the distribution and range of shrubland may spread with increasingly arid conditions. The effects of climate change scenarios on ecosystem functioning may contribute to: (1) the increase of forest decline events; (2) the change in migration patterns; (3) the increase in forest fire risk; (4) the spread of alien species that thrive in the new conditions; (5) the impact on biodiversity with a significant loss of vulnerable species; and (6) the increase of land degradation and soil erosion. Morocco has some general policies to address desertification but without the degree of development and implementation of the Spanish policies.

5.2.3 Vulnerabilities in Socio-economic Sectors

5.2.3.1 Agricultural Vulnerabilities

In the IBRM, unfavourable weather conditions are the greatest risk to subsistence farming systems, especially on marginal land where dry years can have important direct effects on farmers' incomes. The intensification of agriculture within and outside the Reserve will increase pressures on water resources and on land. National policies in both Morocco and Spain are pushing to increase irrigation (and reduce reliance on rain-fed agriculture) and are increasingly focused on water-saving techniques such as drip irrigation, both of which would help to reduce the vulnerability of groups reliant on agriculture for income or subsistence (Le Matin 2011; Inventario de Regadíos 2010). The **Spanish side of the IBRM** has lower levels of agricultural activity than the Moroccan side, but a high proportion of this is irrigated. Recurrent droughts have led to serious damages and loss in the arable sector (cereals, vineyards, and olive trees) and, though to a lesser degree, the pastoral sector. Arboriculture that depends on irrigation in the region is expected to be reduced in the future. In general terms, agricultural water demand is being reduced through increasing the level of more rational and efficient technologies for irrigation. Andalusia has highlighted the need for greater economic diversification in the region to counteract potential economic losses from climate change effects on agriculture (COM 2014). On the **Moroccan side of the IBRM**, agriculture and livestock cover a high proportion of land (60%) and are the most exposed and economically-sensitive sectors to climate change, as traditional rain-fed, extensive farming practices are still the norm, and the main means of subsistence (FAO 2011). This higher risk of loss in productivity ranges from partial to total loss in case of extreme events i.e. droughts and flash-floods. Moroccan farmers have adopted unsustainable agricultural practices (overgrazing, illegal cannabis cropping in higher gradient slopes, forest overexploitation) that have high environmental impacts. In the last decades migration away from these environmentally depleted areas to coastal cities is evident (Barrow and Hicham 2000; Bennis and Sadeq 1998). For the Moroccan population, the secured access to food is a major factor influencing migration as food shortage is one of the main drivers of human migration in the region. At the national level, the Plan Maroc Vert (PMV) has been strongly promoted with a determined focus on irrigation and growth for export purposes. Impacts of this plan are expected to increase pressures on water resources and natural systems in areas around the IBRM. Climate change and reduced water supply threaten to jeopardise the gains originally expected from the PMV and there are now attempts to 'climate proof' the policy. Furthermore, a lack of planning and adequate management of water resources for agriculture means that Morocco will continue to depend on costly imports of cereals in times of drought (World Bank 2010).

5.2.3.2   Tourism

Tourism is an important sector in the IBRM and is very sensitive to climate-related impacts especially during the summer period. Slight increases in mean temperature will favour the extension of the mountain and rural tourism season in the Reserve. However, there are concerns about the effects that higher temperatures may have on the suitability of the region for tourism. There are also worries about the impact of likely water shortages on this part of the economy. Furthermore it is very likely that due to the greater attractiveness of the coast for tourism, the population will migrate from rural areas to urban zones with a consequent loss of local knowledge and identity. **Spain** is the world's third most popular tourist destination in terms of arrivals (UNWTO 2015). However projected changes in climatic conditions will result in a decrease in the Tourist Climatic Index in the 50–60 year timeframe (Ciscar et al. 2009) and it is expected that tourism will be diverted to other parts of Europe during the summer season. Although a reduction in tourism in the Spanish side of the IBRM will have the positive effect of decreasing water demand, this will have a negative impact on the local economy. Spain has made broad policy statements that recognise these coming changes, yet there is little evidence of practical adaptation of the tourism sector. In **Morocco**, tourism in coastal areas can provide alternative employment for rural populations migrating from drought and water scarcity. However, when this migration becomes more permanent, it may lead to the loss of local knowledge, identity and traditional uses of the environment within the IBRM. Morocco's present tourism strategy includes a focus on increasing rural tourism with the aim of decreasing poverty and contributing to local employment (other than agriculture) which may help to reduce migratory flows away from rural areas. It also aims to reward responsible behaviour through

Table 5.1 Summary and estimation of the level of vulnerability for each area of the IBRM (red: high vulnerability, yellow: medium vulnerability, green: low vulnerability)

		IBRM - Spanish side		IBRM - Moroccan side	
		Summary	Vulnerability level	Summary	Vulnerability level
Physical environment	Climate	High annual variability in precipitation and frequent drought with increasing frequency of extreme events. Particular impacts on rural populations directly dependent on resources. Positive and negative effects on energy use in IBRM. Overall good policy framework with cross-cutting strategies. Some challenges remain in terms of implementation.	(yellow)	High annual variability in precipitation and frequent drought with increasing frequency of extreme events. Particular impacts on rural populations directly dependent on resources. Positive and negative effects on energy use in IBRM. Lack of detailed and strategic plan for adaptation to climate change is a major obstacle to addressing impacts.	(red)
	Extreme events – floods	Increased risk of flash-floods with impacts on coastal regions beyond IBRM. Comprehensive policy framework with focus on prevention, protection and preparedness. Need to ensure sufficient attention to flash floods to ensure adaptive capacity.	(green)	Increased risk of flash-floods with impacts on coastal regions beyond IBRM. Lack of forest cover and inappropriate agricultural practices and overgrazing may exacerbate their impacts. Changes in precipitation patterns are seen as problematic. Some experience of flash flooding, flood early warning system in place, but the capacity to mitigate risks is not yet clear.	(yellow)
	Extreme events – droughts	Expected increase in aridity and water scarcity. Spain has moved from crisis-centred approach to drought to increased planning and prevention. Spain's approach has been highlighted at European level as an example of good practice.	(green)	Expected increase in aridity and water scarcity. Historical experience in autonomous adaptation. Disproportional level of vulnerability to drought and water scarcity due to high importance of agriculture to economy and human security. Some systems put in place for drought forecasting and prediction but there is a lack of generalised framework for planning and early warning.	(red)
	Water resources	Existing water deficit, to be exacerbated under climate change. Increased competition for scarce resources due to tourism (e.g. golf courses) and intensively irrigated agriculture outside the IBRM. Integrated water resources management in place. However, need for greater citizen participation and awareness of the need to make use of alternative water sources and prioritisation of sectoral water use according to the realities of climate change.	(yellow)	Existing water deficit, to be exacerbated under climate change. Increasing demand from growth of tourism and agricultural sectors. Water quality and quantity is extremely compromised due to ineffective policies or lack of implementation and monitoring, including stakeholder involvement in management, inadequate use of alternative water sources, lack of controls on discharge of untreated waters, promotion of luxury tourism and high value crop production for export.	(red)

Table 5.1 (continued)

Society and economy	**Forests and natural systems**	Natural systems cover 70% of land on Spanish side of IBRM. Potential climate change impacts include reduction of cork production, increased fire risk, alien species and loss of biodiversity. Higher potential vulnerability of forests than shrublands, climate impacts may produce expansion of shrubland. 70% of the Spanish side of the reserve is protected area and Andalusia has a long-standing and well-developed forest protection policy.	🟩	Natural systems cover 39% of land on Moroccan side of IBRM. Significant impacts expected for ecosystems and soil with loss of native species, increased alien species, risk of forest fires and soil erosion. Moroccan forests in particular are more sensitive to future aridity conditions. Some general desertification policies exist, and there is some effort to protect the forests but not sufficient to address problems of overgrazing and deforestation taking place. Only 30% of the Moroccan side of the Reserve is protected area.	🟨
	Agriculture	Cropland covers around 29% of the Spanish side of the IBRM. Arboriculture that depends on irrigation in the region is expected to be reduced in the future. Pastures are rainfed and no intensification of agriculture (and thus use of water resources from this sector) is expected within the Reserve. The expected use of more rational and efficient technologies for irrigation will reduce water demand in the region.	🟨	Agriculture is a very important economic sector and cropland covers more than 60% of the Moroccan side of the IBRM. The intensification of agriculture outside the Reserve will increase pressures on water resources. Land degradation is high due to overexplotation of land as a result of illegal cannabis plantations, overgrazing and logging. Increase in agriculture for export under the PMV may intensify water use around the IBRM and increase pressures on water and natural systems within and outside the Reserve.	🟥
	Tourism	Economically important sector highly sensitive to climate impacts. Expected reduction in level of tourism as visitors seek locations with less extreme climates. This will reduce water use, but could increase socio-economic vulnerabilities if sector declines. Lack of adaptive strategy in place.	🟥	Current strategy to dramatically increase tourism provides alternative employment in the reserve. However, policy will need to be monitored and consider changing climatic conditions in the regional management plans to ensure that the development of luxury, water dependent facilities do not place additional pressures on non-renewable water resources, and increasing vulnerability.	🟨
	Overall	Overall impacts on natural resources and socio-economic sectors could be substantial, however good policy framework including cross-cutting strategies reduces vulnerabilities. Clear shift in recent decades from a crisis centred approach towards increased prevention and planning of extreme events. Some challenges remain e.g. need for an adaptive strategy for the tourist sector and for effective engagement with stakeholders for adapted water resource management.	🟨	Physical and socio-economic impacts are already felt and set to increase. High dependence on agriculture, lack of protection of natural systems and implementation of sustainable water resource management create particular vulnerabilities and threaten human security in the Moroccan part of the IBRM. Lack of a strategic plan for adaptation to global change which limits the capacity to mitigate the impacts on human populations. Insufficient consideration of cross-cutting effects between sectoral policies.	🟥

the use of 'eco-labels' for environmentally friendly tourism. Nevertheless, the strategy also aims to double the capacity for receiving tourists by 2020 and to triple domestic tourism, including the development of water intensive, irrigation-dependent luxury tourist facilities such as golf resorts. Although it is foreseen that this can be carried out through alternative water sources such as wastewater reuse, if monitoring and implementation of the strategy is inadequate, as has been the case for other water policies, this expansion strategy has the potential to seriously jeopardise water security and livelihoods in the area.

5.2.4   Summary of Vulnerabilities in the IBRM

This chapter has provided a detailed overview of vulnerabilities in the IBRMs physical environment and key socio-economic sectors. Table 5.1 presents a summary and an estimation of the level of vulnerability for each side of the IBRM ranging from high (red) to low vulnerability (green).

5.3   Conclusions

The integrated vulnerability assessment developed and applied in this book encompasses both physical and socio-economic vulnerabilities in the IBRM. Both the Spanish and the Moroccan sides of the Reserve are already subjected to greater climatic variability, reduced water availability, and more frequent extreme events such as flash-floods and droughts, all of which are set to increase in future. Overexploitation of water and other natural resources for socio-economic sectors such as tourism and irrigated agriculture, both inside and around the IBRM, place further pressures on increasingly scarce resources. The results of this assessment demonstrate that although climate change effects are similar for both the Spanish and the Moroccan sides of the IBRM, the impacts will be felt in very different ways. This is in part due to divergent patterns of land-use and socio-economic contexts, but also due to the differing capacity of their policy frameworks to address and adapt to climate change. The Spanish side of the Reserve has been given an overall vulnerability rating of 'medium'. This is due to the existence of policy frameworks that include cross-cutting and inter-sectoral strategies and which are broadly appropriate for addressing the impacts of climate change. Challenges do nonetheless remain in terms of the need to develop an adaptive strategy for the tourist sector and to ensure effective engagement with stakeholders for adapted water resource management. The Moroccan side of the Reserve on the other hand has been given a 'high' overall vulnerability rating. This is primarily due to its heavy reliance on threatened natural systems and water resources to sustain its economically important agricultural sector. The lack of implementation of—at times ill-conceived—water policies, coupled with a lack of protection for natural systems or a strategic plan for climate adaptation create major vulnerabilities and threats to human security on the Moroccan side of the IBRM.

Chapter 6
Conclusions and Recommendations

Abstract This book has analysed the vulnerability of the IBRM to the effects of global and climate change and risks to human security on the basis of a range of inputs including hydro-ecological modelling, water management and climate change scenarios, desk-based research and expert stakeholders. In this final chapter the key research findings are used to develop general insights on effective governance and propose recommendations for action that aim at reducing vulnerability and ensuring human security in the region. The recommendations are organised around the main structural policies. Concerning the water sector, the suggestions highlight the need of ensuring effective mechanisms to prioritise the allocation of water resources and to implement water saving and taxation measures in both sides of the reserve. At the same time, and at national level, increasing public awareness of alternative water resources and engaging with local communities to use them is essential to their acceptance and effective use. In addition to water specific measures, the pursuing of environmentally sensitive sectoral strategies, such as the use of adapted species in the forestry and agricultural sectors, and the implementation of alternative tourism, namely, cultural and eco-tourism, would ensure functional ecosystems while supporting rural economy. Lastly, and in order to ensure equal trans-boundary benefits for the whole IBRM, a practical joint action plan covering the entire reserve needs to be put in place and effectively implemented.

Keywords Mediterranean basin · Spain · Morocco · Climate change · Climate adaptation · Adaptive capacity · Human security · Food security · Water security · Agriculture · Migration · Floods · Water scarcity · Drought · Policy frameworks · Vulnerability

6.1 Conclusions

The preceding chapters have outlined the water-related impacts of climate change, vulnerabilities and risks to human security in the protected area of the Intercontinental Mediterranean Biosphere Reserve (IBRM) shared by Spain and

© The Author(s) 2017   77
D. Abdul Malak et al., *Adapting to Climate Change*,
SpringerBriefs in Environmental Science, DOI 10.1007/978-3-319-51680-6_6

Morocco. The IBRM and the Mediterranean region in general are already experiencing increasingly severe climate change effects such as higher average temperatures as well as extreme events such as unprecedented heat waves, severe droughts, and major floods. These impacts threaten environment, society and economy and can place human security at risk. However, the level of vulnerability and consequences for human security vary greatly. These are determined both by the severity of the physical and socio-economic impacts and the extent to which socio-ecological systems can be protected or adapted to change through appropriate policy frameworks. This book has developed and conducted an integrated vulnerability assessment that combines hydro-ecological, socio-economic and policy assessments. On this basis, the impacts and adaptive capacity of the IBRM are described and conclusions drawn on the vulnerability of the Reserve, distinguishing between the Spanish and Moroccan sides of it. In this final chapter key research insights are brought together accompanied by recommendations for reducing vulnerability and ensuring human security within the IBRM and the region more generally.

From the point of view of its biophysical conditions, the IBRM is relatively similar on both the Spanish and the Moroccan sides. This includes a similar context of natural water scarcity and experience of extensive and recurrent droughts of increasing severity e.g. during the mid-1990s. Flash floods have a direct impact on soil erosion and significant soil losses have been registered in both northern Morocco and southern Spain (Ouassou et al. 2005; RME 2008). Climate change is expected to exacerbate the decreasing level of freshwater availability and reduced soil productivity as well as biodiversity loss in the IBRM with associated impacts on ecosystems and their services (Garcia et al. 2010; Magnan et al. 2009). Rapid socio-economic growth and technological transformation in and around the Reserve in both countries (e.g. urban expansion on the Costa del Sol in Spain and the Tanger-Med port in Morocco) have increased pressures on natural resources. Unsustainable development coupled with the paucity of structural water resources and the initial impacts of climate change have led to overexploitation of water resources across both sides of the IBRM with considerable consequences for the hydrological cycle. This has knock-on effects for the sustainability of sectors such as tourism; reduced water supply for resorts and water shortages in many villages coupled with uncomfortable summer temperatures may reduce the future desirability of the region as a holiday destination (Méndez 2008) if no measures are taken to adapt.

The IBRM is on a key migratory route to Europe, which itself has the highest global net level of international migration (UNDESA 2013). This leads to additional pressures on society, economy and infrastructure, in particular on the Moroccan side of the Reserve. The majority of migration is due to (civil) protection needs (Banulescu-Bogdan and Fratzke 2015) but the effects of global and climate change such as increased incidence of extreme weather events, are increasing displacement and migration (IOM 2009). During dry periods, there is also movement within the IBRM from rural to urban centres on both the Spanish and Moroccan sides. This migratory trend is particularly intense during the region's frequent poor harvests which damage the local agricultural economy (Refass 1992; CRED 2010, 2014) and precipitate rural-urban migration by individuals in pursuit of economic security (employment) and water security (access to drinking water) (Personal communication, stakeholders' meeting 2012). Although the complexity of factors that lead to migration means that there are no reliable estimates on numbers, in general, migratory trends are expected to increase as a possible consequence of climate change (ENPI 2007; Fermin 2009a, b; IOM 2009). Such autonomous adaptation can lead to a loss of traditional knowledge related to rural areas and indeed rural-urban migration to coastal cities may even create new vulnerabilities by exposing higher numbers of the population to coastal floods and sea-level rise.

Despite some homogeneity in the physical context and socio-economic dynamics, there are important differences between the Spanish and Moroccan sides of the IBRM which alter their respective vulnerabilities. First of all, land-use and socio-economic context differ considerably between the two sides of the Reserve. On the Spanish side, more than half of the territory of the IBRM is covered by forest, scrubland and grassland (70%); and around 29% of cropland, with 70% designated as protected area. Conversely on the Moroccan side, only 39% of the territory is covered by forest, scrubland and grassland, with more than 60% dedicated to crops and a mere 30% as protected area (GlobCover v.2.3 2009, 2010). The high percentage of cropland in Morocco reflects the central role that agriculture plays in the national (and local) economy with nearly half the country's population employed in this sector. The differences in land-use mean that impacts will be more significant for forests and natural systems in Spain and for cropland in Morocco.

Secondly and most importantly are the policy frameworks on the two sides of the Reserve. It is above all the governance systems, institutions, policies and laws, as well as their implementation that determine the final vulnerability of the IRBM. As an industrialised nation, Spain has a greater overall capacity to address natural and climate change induced vulnerabilities. Also, as a Member State of the European Union, it has complied with the environmental *acquis*. This has meant a minimum level of environmental policy development, implementation and monitoring. For example Spain has comprehensive and sustainable approaches to drought and flood risk management in place that in some cases have already shown their capacity to reduce vulnerabilities. Adaptation to climate change has also been addressed through a national strategy and integrated as a cross-cutting issue in a range of policies and laws. In addition to this basic compliance, Spain and the Autonomous Community of Andalusia have acted as front-runners in drought management and forest protection policies respectively. Spain could however do more in terms of developing a dedicated adaptation strategy for the tourist sector and to engage further with stakeholders for effective and equitable water management in the context of climate change.

From the Moroccan side, there are some efforts to reduce environmental and socio-economic vulnerabilities. However, on the whole, these policies are broad-stroke strategic planning documents with implementation that is slow or lacking and scarce engagement with stakeholders. Water quality and quantity are seriously compromised by this inadequate implementation and monitoring of policies (e.g. adequate treatment and discharge of wastewater) as well as the creation of 'maladaptive' policies (e.g. rapid dam building leading to siltation). There is a heavy reliance on—mostly rain-fed—agriculture as a primary source of food and economic security in the Reserve. This makes the local population particularly vulnerable to climate change impacts. In order to adapt, and in the absence of a national adaptation strategy, the population pursues environmentally harmful autonomous strategies such as unsustainable agricultural practices (overgrazing, illegal cannabis cropping in higher gradient slopes, forest overexploitation) to secure livelihoods and ensure human security. Finally, there is often insufficient consideration of coherence between sectoral policies. For example, the Plan Maroc Vert aims to intensify agricultural production for export; while the Vision 2020 Strategy for tourism aims to double tourism by 2020. Without due consideration of potentially competing objectives, and clear guidelines on the use of alternative, renewable water resources, these strategies could lead to further overuse of scarce resources. Despite these shortcomings, it must be underlined that the difference between Spanish and Moroccan GDP and human development is substantial with understandable implications for the capacity to implement environmental governance.

6.2   Recommendations

This book has analysed the vulnerability of the IBRM to the effects of global and climate change and risks to human security on the basis of a range of inputs (hydro-ecological model, water management and climate change scenarios, desk-based research and expert stakeholders). The assessments (hydro-ecological, socio-economic, policy and vulnerability) conducted have brought to light a number of issues that need to be addressed in the IBRM. This final section summarises key findings and general insights on effective governance and proposes recommendations for action to reduce vulnerability and risks to human security.

Water policy

(1) **Increase awareness and uptake of alternative water resources and water saving measures**

In the face of a more arid environment with reduced water resources, there is a pressing need for infrastructures that support the use of unconventional and renewable water resources (e.g. wastewater reuse) especially on the Moroccan side of the IBRM. Such solutions reduce pressures on renewable water resources and are already beginning to be put into practice in the tourist sector for the irrigation of golf courses. At the same time, and at the national level, increasing public awareness of alternative water resources and engaging with the local communities to use them are essential to their acceptance and effective use. In parallel to the promotion of unconventional water resources, water saving measures such as drip irrigation or shower adaptors need to be further introduced for domestic and industrial use, with a special emphasis on the tourism and agricultural sectors.

(2) **Improve and monitor water quality**

Increasingly scarce rainfall, overuse of water resources and pollution are of major concern for water quality. There is a need to set and monitor water quality standards, in particular on the Moroccan side of the IBRM, where urgent actions are required in order to control the disposal of untreated water and dumping of waste to reduce contamination of groundwater. Both inside and outside the Reserve, a reduction in intensive farming, or relocation to less environmentally sensitive areas could improve water quality by reducing diffuse pollution.

(3) **(Re-)prioritise the allocation of water resources**

In both Spain and Morocco, stakeholders highlighted the need for a flexible approach to the management and allocation of water resources that can be adapted to changing realities and needs. To ensure human security, water allocation should be prioritised according to human needs especially during drought episodes, ensuring domestic water supply as a priority, followed by agriculture for local food security. Allocation of the remaining resources to other activities of economic importance such as tourism or export agriculture should follow only where options to use unconventional resources have first been exhausted.

(4) **Develop effective pricing and taxation of water resources**

Full cost recovery and effective water pricing are areas which could be further developed in both the Spanish and Moroccan policy contexts particularly in relation to wastewater treatment. Sectoral water taxation—particularly in tourism or agriculture—and effective pricing would help to regulate water consumption. In Morocco, there is a social demand to reinvest the economic benefits of water sector in the conservation and protection of natural resources.

Other sectoral policies

(5) **Pursue environmentally sensitive sectoral strategies**

While sectoral policies in the IBRM must aim to create socio-economic value, it is essential that this is done without negatively affecting the healthy functioning of land-based and aquatic ecosystems. Sectoral strategies in the IBRM should protect natural systems and water resources in order to ensure the sustainability of the sectors as well as reduce vulnerability and risks to human security.

- For forestry, effective management options include the replacement of highly flammable species, water stress reduction, regulation of age-class distributions and widespread management of accumulated fuel. Governments can provide sectoral support through the development of advanced systems of forest inventories and forest health monitoring.
- For agriculture, the reintroduction of traditional crop species and varieties that are adapted to the local climate—including frequent drought conditions—in

addition to changes to agricultural schedule and agronomic practices will reduce losses and support the shift to low water consumption irrigation systems.

- For tourism, the promotion of new forms of tourism such as eco-tourism or cultural tourism will favour the recharge of aquifers and positively influence the preservation and/or restoration of regional ecosystems.

(6) Mainstream adaptation across sectors

To reduce vulnerabilities to global and climate change all sectors must develop measures to adapt. However, to avoid 'maladaptation', cross-sectoral coordination of these adaptive measures is essential. Spain's adaptation strategy gives detailed consideration to vulnerabilities and human security and Spain has established numerous coordination mechanisms for mainstreaming climate adaptation across different sectors. This approach could act as a model in the IBRM to ensure effective adaptation of sectors active in the Reserve.

(7) Develop policies to support the rural economy

The rural populations living in the reserve are particularly vulnerable to climate impacts, especially where they are directly reliant on agriculture for economic and food security. In times of extreme events such as drought or flash flooding, temporary migration leads to excessive pressures on urban centres. In order to reduce this internal displacement, and to reduce vulnerabilities, policies are needed to support the human security of rural populations. Ensuring reliable household water supply in times of drought is an essential component of this. Furthermore, policies for economic diversification e.g. eco-tourism or production of artisanal goods would lead to greater economic security and avoid the need to plant erosion-inducing cash crops such as cannabis.

Engagement and cooperation

(8) Strengthen practical cooperation

The creation of the IBRM is an opportunity for cooperation between Spain and Morocco and for supporting sustainable development in the region as a whole. A broad cooperation strategy exists, but to date there has been little in the way of practical joint action. Practical cooperation necessitates a strengthening of common institutions to implement coherent measures to the benefit of the entire Reserve and the allocation of a corresponding budget. Reserve managers from both sides could participate in further knowledge sharing and training exercises to establish a common baseline of information and a joint approach for the effective management of the IBRM.

(9) Improve monitoring and information sharing on impacts and vulnerabilities

This book provides a comprehensive account of existing and potential vulnerabilities in the IBRM and has highlighted areas for concern. Monitoring of selected environmental and socio-economic indicators would ensure that the understanding of impacts and vulnerabilities in the IBRM is built upon and updated. Establishing systems for information-exchange between Spain and Morocco would further strengthen this knowledge base and would benefit communication and cooperation in the management of the Reserve more generally. Indicators could include environmental and socio-economic data e.g. on climate, natural systems, water quality and quantity, migration and demographics.

(10)   **Increase civic engagement**

Empowering the local population to participate in the management and planning of the IBRM can increase the sense of individual responsibility for the protection and use of natural resources. The involvement of inhabitants also allows for the consideration and integration of relevant local knowledge (for example on locally adapted crop varieties, water storage or irrigation) and indigenous governance systems (for example water controls in times of water scarcity).

Annex A
Participant Institutions in the Stakeholders Meetings and Focus Group

Andalusian Stakeholders meetings	Moroccan Stakeholders meetings
Administration of the Natural Park "*Sierra de Grazalema*"	Administration of the National Park "*Talassemtane*"
Water New Technologies Centre "*Centro de Nuevas Tecnologías del Agua*" (CENTA) (2 participants)	Administration of the Water and forest Regional Direction "*Direction régional d'Eau et Forets*" (4 participants)
Hydro-geological Centre of the Malaga University "*Centro de Hidrogeología de la Universidad de Málaga*" (CEHUIMA)	Administration of the Agriculture Regional Direction "*Direction Provinciale d'Agriculture*"
Ecological Department of the Jaen University "*Departamento de ecología, Universidad de Jaén*"	Local development NGO of Chefchaouen "*Association de Développement local de Chefchaouen (ADL)*" (2 participants)
Spanish Oceanography Institution "*Instituto Español de Oceanografía*"	Environmental and development NGO of Talassemtane "*Association Talassemtane de l'Environnement et de Développement (ATED)*"
Andalusian Focus group	Moroccan Focus group
Administration of the Natural Park "*Sierra de Grazalema*"	Administration of the National Park "*Talassemtane*"
Administration of the Protected Naturals Spaces in Malaga	Administration of the Tourism Delegation of Tetuan
Hydro-geological Centre of the Malaga University "*Centro de Hidrogeología de la Universidad de Málaga*" (CEHUIMA)	Administration of the Water and forest Regional Direction (2 participants)
Touristic Initiatives Centre "*Centro de Iniciativas Turísticas*" (*CIT*)	Granada University and Abdelmalek Essaadi University
Water Culture New Foundation "*Fundación Nueva Cultura del Agua*"	Environmental and development NGO of Talassemtane "*Association Talassemtane de l'Environnement et de Développement (ATED)*"
Water New Technologies Centre "*Centro de Nuevas Tecnologías del Agua*" (CENTA)	Local development NGO of Chefchaouen "*Association de Développement local de Chefchaouen (ADL)*"
"*Los Alcornocales*" rural development NGO "*Grupo de desarrollo rural Los Alcornocales*"	WWF MedPo Environmental NGO
	TARGA—Aide Development NGO

© The Author(s) 2017
D. Abdul Malak et al., *Adapting to Climate Change*,
SpringerBriefs in Environmental Science, DOI 10.1007/978-3-319-51680-6

Annex B
Overview of Discussions from the Focus Group Meetings

Scenario A—Morocco

Adaptation measures	Actors	Means	Barriers and opportunities
Civil society awareness and engagement	The administration is responsible for the implementation of measures	Taxes on water consumption	B—Exhaustion of water resources for the intensification of export crops
Integration of local knowledge and promotion of local products and activities:	Civil society (public awareness and participation and integration of civil society actions in policies)	Sensitisation, public awareness campaigns	B—Salt-water intrusion
Regional agricultural products such as herbs, oil, jam, honey		Implementation of existing policies	B—Cumulative impacts of drought
Handicrafts		New laws	B—Siltation of dams
Promotion of rural and sustainable tourism (bicycle, 'bed and breakfast')		Promotion of more efficient approaches and technologies	B—Reduction in public subsidies
Cooperatives association			B—Morocco has signed all the international conventions but the application of policies is still poor
Return to traditional values and local knowledge			O—Knowledge of agricultural and forestry varieties that are better adapted to climate change
Reforestation			O—Public involvement in decision-making
Restoration and conservation of biodiversity			
Maintain health of natural systems			
Promotion of organic agriculture			

(continued)

© The Author(s) 2017
D. Abdul Malak et al., *Adapting to Climate Change*,
SpringerBriefs in Environmental Science, DOI 10.1007/978-3-319-51680-6

(continued)

Adaptation measures	Actors	Means	Barriers and opportunities
Reduction in water use			
Improvements to research on water resources			
Impact evaluation for new projects. The current law already demands this impact evaluation. There is however a gap between the person who makes the impact evaluation and who will put it into practice			
Increase economic support to promote adaptation measures. There are already subsidies for improving irrigation systems (drip-fed irrigation). This subsidy should be expanded			
Changes to agricultural and forestry varieties for species that consume less water			

Scenario D—Morocco

Adaptation measures	Actors	Means	Barriers and opportunities
Desalination Dam building (already in place) Wastewater and residual water reuse Improvement of water infrastructures	The administration is responsible for establishing measures Civil society	Awareness raising Research Link between research, civil society and the administration	B—Raised costs of desalination (energy consumption)

(continued)

(continued)

Adaptation measures	Actors	Means	Barriers and opportunities
Development of new technologies Improvement of irrigation efficiency Promotion of water saving (reduction of consumption) Increase density of housing in urban areas so as to reduce water demand from e.g. individual gardens and swimming pools Reforestation using adapted species for agriculture and forestry Promotion of research into species in danger of disappearance Adapt existing agricultural calendar (used to plan the planting and harvesting of different crops) to account for climate change Return to traditional practices and local knowledge (autonomy of civil society) Legalisation of cannabis for medical use Control external demand for cannabis		Reinforcement of the implementation of existing policies Taxes for the use of resources Recuperation of traditional and local knowledge Reinforcement of cannabis prosecutions	

Scenario A—Spain

Measures	Actors	Means	Barriers and opportunities
Integrated use of water Fulfilment of the regulations in force (hydrological plans, water framework directive, Water Law etc.) water treatment and reuse of rain water	Administration Civil society through public participation (there is no tradition of citizen participation and it generates a lack of interest due to the final decisions being political)	Reinforce existing legislation Compliance with existing legislation	B—Existing regulations are very strict O—Make the challenges positive O—Waste water re-use

(continued)

(continued)

Measures	Actors	Means	Barriers and opportunities
Improvement of infrastructures and water saving measures to increase the availability of the resource Artificial recharge of groundwater Use of alternative sources of water (water reuse) Improvement of the quality of the resource which would improve the state of the ecosystems Increase in the capacity of the water regulation and storage Improve Research, Development and innovation in early warning systems, regeneration and reuse of waters Upgrade and improve tourism from a water perspective, which is currently lacking Tourist taxes to be reinvested in the Reserve Appropriate tariffs on water Create municipal emergency plans (at town council level). Set up immediate early warning systems Emergency management rather than disaster management Plans for reforestation of river basins, maintenance and follow-up	The Water Framework Directive promotes consultations, not participatory processes Co-responsibility of citizens in water management	Increase Monitoring and control of water resources Capacity building for citizens Planning	O—The expected changes can help to plan now and to improve the future situation

(continued)

Measures	Actors	Means	Barriers and opportunities
Improve the social perception of reused waters			
Energy efficiency (technical building code)			

Scenario D—Spain

Measures	Actors	Means	Barriers and opportunities
Flexibility in the mechanisms for reassigning water resources (in accordance with the law and socio-economic optimisation)	Water administration with citizen co-responsibility through participation	Enforce implementation of the law	
Application of international, national and regional legislative framework for water and energy (control and monitoring)	Europe promoting more research	Encourage water savings, infrastructure and changes in water use towards other more cost-effective uses with less consumption	
Create a public hydric authority watchdog that would ensure the fulfilment of the law		Tourist tax that can be reinvested in water	
Development of agricultural policy to provide farmers with knowledge and tools to adapt to climate impacts		Water tariffs	
		Research, development and innovation to supplement the lack of information	
Increase food production to ensure that the increasing food demand of the touristic sector is met while still ensuring food security for the rest of the population		Agricultural policies to incentivise particular types of crops that are better adapted to the climate	
		Preempting of the situation	

(continued)

(continued)

Measures	Actors	Means	Barriers and opportunities
Increased efficiency of water use in agriculture through irrigation technologies			
Reprioritise water use away from urban and touristic use towards agriculture in times of drought			
Regulation of water allocation in accordance with demand. Demand-side management			
Water re-use (take full advantage of treated water)			
New systems for water storage			
Citizen education in the sustainable use of resources, use of alternative and clean energy sources and conservation			
Price tariffs			
Sustainable tourism			

Bibliography

AAA, Agencia Andaluza del Agua (2011) Plan Hidrológico de la Demarcación Hidrográfica de las Cuencas Mediterráneas Andaluzas. Anejo 2: Recursos hídricos., Sevilla: Junta de Andalucia

Abdesslam B (2011) Social determinants and health equity in Morocco. s.n., Rio de Janeiro, pp 1–22

Adger N (2003) Social aspects of adaptive capacity. In: Smith J, Klein R, Huq S (eds) Climate change, adaptive capacity and development. Imperial College Press, London, s.l.

AEMET (2009) Generación de escenarios regionalizados de cambio climático para España. s.n., s.l.

Agoumi A (2003) Vulnerability of North African countries to climate change. Climate Change Knowledge Network, s.l., pp 1–11

Albiac J et al (2006) The rise and fall of the Ebro water transfer. Nat Res J 46(3):727–757

Alpert P et al (2002) The paradoxical increase of Mediterranean extreme daily rainfall in spite of decrease in total values. Geophys Res Lett 29(1536)

Amelung B, Viner D (2006) Mediterranean tourism: exploring the future with the tourism climatic index. J Sustain Tourism 14(4):349–366

Ameziane T, Belghiti M, Benbeniste S, Bergaoui M et al (2001) Drought management guidelines: examples of application in Mediterranean countries. In: Iglesias A et al (eds) European Commission—EuropeAid Co-operation Office, Euro-Mediterranean Regional Programme for Local Water Management (MEDA Water) and Mediterranean Drought Preparedness and Mitigation Planning (MEDROPLAN), s.l.

Anon (2016) The National Parks Autonomous Agency (OAPN) of the Ministry of Agriculture, Food and Environment. s.n., Lima

Arnold J, Srinivasan R, Muttiah R, Williams J (1998) Large-area hydrologic modeling and assessment: Part I. Model development. J Am Water Resour Assoc 34(1):73–89

Banulescu-Bogdan N, Fratzke S (2015) Migration information source. s.n., s.l.

Barrera-Escoda A, Llasat M (2015) Evolving flood patterns in a Mediterranean region (1301–2012) and climatic factors—the case of Catalonia. Hydrol Earth Syst Sci 19:465–483

Barrow C, Hicham H (2000) Two complementary and integrated land uses of the Western High Atlas Mountains, Morocco: the potential for sustainable rural livelihoods. Appl Geogr 20 (4):369–394

BCDC, San Francisco Bay Conservation and Development Commission (2011) Adapting to Rising Tides—Assessing Climate Change Vulnerability & Risk, Staff Report. http://www.bcdc.ca.gov, Dec 2011, s.l.

Bekkari L, Kadiri Z, Faysse N (2008) Appropriations du cadre de l'association des usagers des eaux agricoles par les irrigants au Maroc. Analyse comparative de cas au Moyen Atlas et Moyen Sebou. Jun 2007. Cirad, En: Nabeul, 8 p

Beniston M (ed) (2013) Regional assessment of climate change in the Mediterranean. In: Advances in global change research series. Springer, s.l., pp 9–29

Bennis A, Sadeq H (1998) Case study: Morocco, population and irrigation water management. s.n., s.l.

© The Author(s) 2017
D. Abdul Malak et al., *Adapting to Climate Change*,
SpringerBriefs in Environmental Science, DOI 10.1007/978-3-319-51680-6

Benoît G, et Comeau A (2005) Méditerranée: les perspectives du Pan Bleu sur l'environnement et le développement. Plan Bleu, s.l.

Beven K (2011) I believe in climate change but how precautionary do we need to be in planning for the future? Hydrol Process 25:1517–1520

Blinda M et al (2006) Technical report on water scarcity and drought management in the Mediterranean and the Water Framework Directive. Med Joint Process WFD/EUWI, s.l.

Bruggeman A, Hadjinicolaou P, Lange M (2010) Climate outlooks for CLICO case study sites. CLICO Climate Change, Hydro-conflicts and Human Security, WP2. s.n., s.l.

Brugha R, Varvasovszky Z (2000) Stakeholder analysis: a review. Health Policy Plan 15 (3):239–246. doi:10.1093/heapol/15.3.239

Bryson J (2003) What to do when stakeholders matter: a guide to stakeholder identification and analysis techniques. Public Management Review. s.n., s.l.

Carrasco F et al (2008) Application of the European water framework directive in a Western Mediterranean basin (Málaga, Spain). Environ Geol 54:575–585

CEDEX (2012) Estudio de los impactos del cambio climático en los recursos hídricos y las masas de agua. Informe final, Tomo único. Informe técnico para el Ministerio de Agricultura, Alimentación y Medio Ambiente, s.l. CEDEX 40-407-1-001

CESE, Conseil Economique, Social et Environnemental (2014) La gouvernance par la gestion intégrée des ressources en eau au Maroc. Levier fondamental de développement durable Auto-Saisine n° 15/2014, s.l.

Choukr-Allah R (2011) Comparative study between Moroccan water strategies and WFD. Options Méditerranéennes, p 98

Ciscar J et al (2009) Climate change impacts in Europe. Final report of the PESETA research project, s.l., Luxembourg, Office for Official Publications of the European Communities

COM (2007) Drought management plan report: including agricultural, drought indicators and climate change aspects. Technical Report 2008—023, s.l. Nov 2007

COM (2014) Spain: Common Agricultural Policy Factsheet. [Online] Available at: http://ec.europa.eu/agriculture/cap-in-your-country/pdf/es_en.pdf. Accessed 5 May 2016

COM (2016) Designing the path—a strategic approach to EU agricultural research & innovation. s.n., Brussels

CRED, Center for Research on the Epidemiology of Disasters (2010) International Disasters Data Base (EM-DAT). s.n., University of Louvain and the United Nations Department of Humanitarian Affairs

CSEC (2014) Plan National de l'Eau. Conseil Supérieur de l'Eau et du Climat. s.n., s.l.

Dessai S, Hulme M (2004) Does climate adaptation policy need probabilities? Clim Policy 4 (2):107–128

Downward S, Taylor R (2007) An assessment of Spain's Programa AGUA and its implications for sustainable water management in the province of Almería, southeast Spain. J Environ Manage 82(2):277–289

EEA, European Environment Agency (2005) Vulnerability and adaptation to climate change in Europe. EEA Technical Report No 7/2005, ISSN 1725-2237. s.n., Copenhagen

EEA, European Environment Agency (2012) Climate change, impacts and vulnerability in Europe 2012; EEA technical Report EEA Report No 12/2012. s.n., Copenhagen

EEA, European Environment Agency (2015) SOER 2015: climate change impacts and adaptation. European Environment Agency, Copenhagen

EFE (2016) Gobierno aprueba 16 planes gestión de inundaciónes, dotados con 800 millones. 15 May 2016 [Online] Available at: http://www.efe.com/efe/english/varios/european-commission-urges-quick-ratification-of-paris-climate-accord/50000269-2911275#

El Alaoui M (2004) Les pratiques participatives des associations d'usagers de l'eau dans la gestion de l'irrigation au Maroc: étude de cas en petite, moyenne et grande hydraulique. Rabat Rabat, du 19 au 23 avril 2004, s.n., pp 146–163

El Kharraz J, El-Sadek A, Ghaffour N, Mino E (2012) Water scarcity and drought in WANA countries. Procedia Eng 33:14–29

El Khatri S, El Hairech T (2014) Drought conditions and management strategies in Morocco. s.n., s.l.

ENPI, European Neighbourhood and Partnership Instrument (2007) Morocco: strategy paper (2007–2013). European Neighbourhood and Partneship Instrument, s.l.

EPA, Environmental Protection Agency (2010) Climate Change Vulnerability Assessments: A Review of Water Utility Practices. Office of Water, EPA 800-R-10-001, s.l., Aug 2010

Estrela T, Vargas E (2012) Drought management plans in the European Union. Water Resour Manage 26:1537–1553

Eurobarometer (2014) Special Eurobarometer Report 410. Europeans, agriculture and the Common Agricultural Policy (CAP). [Online] Available at: http://ec.europa.eu/public_opinion/archives/ebs/ebs_410_en.pdf. Accessed 5 May 2016

Ewen J, O'Donnell G, Burton A, O'Connell E (2006) Errors and uncertainty in physically-based rainfall-runoff modelling of catchment change effects. J Hydrol 330:641–650

FAO (2011) Climate change, water and food security. FAO water reports, s.l., p 36

FAO (2015) Water for food security and nutrition. The High Level Panel of Experts on Food Security and Nutrition, Rome

Fermin A (2009a) Case study report for the EACH-FOR project: Morocco

Fermin A (2009b) Case study report for the EACH-FOR project: Spain. s.n., s.l.

Filipe A, Lawrence J, Bonada N (2013) Vulnerability of stream biota to climate change in Mediterranean climate regions: a synthesis of ecological responses and conservation challenges. Hydrobiologia 719(1):331–351

Font N, Subirats J (2010) Water management in Spain: the role of policy entrepreneurs in shaping change. Ecol Soc 15(2):25

Fornés J, de la Hera A, Llamas R (2005) The silent revolution in groundwater intensive use and its influence in Spain. Water Policy, 1–16

Fussel H (2007) Vulnerability: a generally applicable conceptual framework for climate change research. Glob Environ Change 17:155–167

Garcia N, Cuttelod A, Abdul Malak D (eds) (2010) The status and distribution of freshwater biodiversity in Northern Africa. IUCN, Gland, Cambridge, and Malaga

Garrido A, Iglesias A (2007) Groundwater's role in managing water scarcity in the Mediterranean Region. s.n., Alicante

Garrido A, Llamas M (2009) Water policy in Spain. Routledge, New York, 234 pp

Gassman P, Reyes M, Green C, Arnold J (2007) The soil and water assessment tool: historical development, applications, and future research directions. Trans ASABE 50(4):1211–1250

Gerstetter et al (2011) Review of international and national policies and institutional frameworks. CLICO project. s.n., s.l.

Gerstetter C, Kampa E, McGlade K, Timeus K (2012) Review of international and national policies and institutional frameworks. CLICO Output, s.l.

Giannakopoulos C et al (2009) Climatic changes and associated impacts in the Mediterranean resulting from a 2 °C global warming. Glob Planet Change 68(3):209–224

Gillette H (1950) A creeping drought under way. Water and Sewage Works, pp 104–105

Giorgi F (2006) Climate change hot-spots. Geophys Res Lett 33(L08707)

GlobCover v.2.3 (2009, 2010) GlobCover Land Cover Maps. [Online] Available at: http://due.esrin.esa.int/page_globcover.php

Government of Morocco (2013) Stratégie du Tourisme Vision 2020. Rabat. Presentation. http://www.icao.int/meetings/moroccan-economic-forum/documents/tourismmaroc.pdf. s.n., s.l.

Government of Morocco (2015) Intended Nationally Determined Contributions (INDCs) Maroc: Contribution prévue déterminée au niveau national (CPDN) dans le cadre de la CCNUCC., Submitted 5 June 2015. Retrieved from http://www4.unfccc.int/submissions/INDC/Submission%20Pages/submissions.aspx. 25 April 2016. s.n.

Gritti E, Smith B, Sykes M (2006) Vulnerability of Mediterranean Basin ecosystems to climate change and invasion by exotic plant species. J Biogeogr 33:145–157

Hammill A, Bizikova L, Dekens J, McCandless M (2013) Comparative analysis of climate change vulnerability assessments: lessons from Tunisia and Indonesia. Published by Deutsche Gesellschaft für Internationale Zusammenarbeit (GIZ) GmbH, Eschborn, s.l., March 2013

HCEFLCD, Haut Commissariat aux Eaux et Forêts et à la Lutte Contre la Désertification (2015) Gestion de la biodiversité. http://www.eauxetforets.gov.ma/fr/text.aspx?id=1026&uid=45. Accessed 07 Dec 2015. s.n., s.l.

Hisdal H, Stahl K, Tallaksen L, Demuth S (2001) Have streamflow droughts in Europe become more severe or frequent? Int J Climatol 21:317–333

Hoff H (2011) Understanding the Nexus. Background Paper for the Bonn 2011 Conference: The Water, Energy and Food Security Nexus. Stockholm Environment Institute (SEI), Stockholm

Howe C (2005) Property rights, water rights and the changing scene in western water. Water institutions: policies, performance and prospects. Springer, Berlin, pp 175–185

IGBP (2015) Earth System definitions—IGBP. [Online] Available at: www.igbp.net. Accessed 8 Sept 2015

Iglesias A, Moneo M (2005) Drought preparedness and mitigation in the Mediterranean: analysis of the Organizations and Institutions. Options Méditerranéennes, Series B, n° 51. Universidad Politécnica de Madrid, Madrid

Iglesias A, Garrote L, Flores F, Moneo M (2007) Challenges to manage the risk of water scarcity and climate change in the Mediterranean. Water Resour Manage 21:775–788

INE (2016) Boletín de estadística. Instituto Nacional de Estadística, s.l.

Inventario de Regadíos (2010) Inventario y caracterización de los regadíos en Andalucía. Colección Agricultura, Serie Riegos. Junta de Andalucía, s.l.

IOM (2009) Climate change, environmental degradation and migration: addressing vulnerabilities and harnessing opportunities. Int Migr J, 104

IPCC (2001) Climate change 2001: impacts, adaptation, and vulnerability. contribution of Working Group II to the third assessment report of the Intergovernmental Panel on Climate Change. Cambridge University Press, Cambridge

IPCC (2002) Climate change and biodiversity: IPCC technical paper V. Intergovernmental Panel on Climate Change, s.l.

IPCC (2007a) Climate Change 2007: Impacts, adaptation and vulnerability. Contribution of Working Group II to the fourth assessment report of the Intergovernmental Panel on Climate Change. Cambridge University Press, Cambridge

IPCC (2007b) Climate change 2007: the physical science basis. In: Solomon S et al (eds) Contribution of Working Group I to the fourth assessment report of the IPCC. Cambridge University Press, Cambridge and New York

IPCC (2012) Managing the risks of extreme events and disasters to advance climate change adaptation. s.n., New York

IPCC (2014) Climate change 2014: synthesis report, IPCC. s.n., Geneva, 151 pp

Jeune Afrique (2016) Le Maroc lance un plan d'urgence contre la sécheresse. 29 Jan 2016 http://www.jeuneafrique.com/298162/economie/maroc-lance-plan-durgence-contre-secheresse/

Jordà G, Marbà N, Duarte C (2012) Mediterranean seagrass vulnerable to regional climate warming. Nat Clim Change 2:821–824

Jump A, Hunt J, Peñuelas J (2006) Rapid climate change-related growth decline at the southern range edge of Fagus sylvatica. Glob Change Biol 12:2163–2174

Kallis G (2010) Case study protocol. Clico project Workpackage 1. Internal report. s.n., s.l.

Karaky R, Arndt C (2002) Climate variability and agricultural policy in Morocco. Purdue University, West Lafayette, s.l.

Karas J (1998) Climate change and the Mediterranean region. Excecutive Summary for Greenpeace. Greenpeace, s.l.

Keenan T, Sabate S, Gracia C (2010) Soil water stress and coupled photosynthesis conductance models: bridging the gap between conflicting reports on the relative roles of stomatal, mesophyll conductance and biochemical limitations to photosynthesis. Agric For Meteorol 150:443–453

Kerr R (2005) Confronting the bogeymon of the climate system. Science 310:432–433

Klausmeyer K, Shaw M (2009) Climate change, habitat loss, protected areas and the climate adaptation potential of species in Mediterranean ecosystems worldwide. PLoS ONE 4(7)

Kostopoulou E et al (2014) Spatio-temporal patterns of recent and future climate extremes in the eastern Mediterranean and Middle East region. Nat Hazards Earth Syst Sci 14:1565–1577

Kundzewicz ZW et al (2013) Flood risk and climate change: global and regional perspectives. Hydrol Sci J 59(1):1–28

Kutsch H (1982) Principle features of a form of water-concentrating culture on small-holdings with special reference to the anti-Atlas. Trierer Geogr. Studien 5, Trier

Larroudé P et al (2014) Three simple indicators of vulnerability to climate change on a Mediterranean beach: a modeling approach. Ocean Eng 76:172–182

Le Matin (2011) Agriculture durable. Lente reconversion au goutte-à-goutte, 17 Apr 2011

Le Plan Maroc Vert (2008) s.n., s.l.

Lenton T et al (2008) Tipping elements in the Earth's climate system. Proc Natl Acad Sci U S A (PNAS) 105(6):1786–1793

Lindner M et al (2010) Climate change impacts, adaptive capacity, and vulnerability of European forest ecosystems. For Ecol Manage 259(4):698–709

Llamas MR (2000) Some lessons learnt during the drought of 1991–1995 in Spain. In: Vogt, Somma (eds) Drought and Drought Mitigation in Europe. Kluwer Academic Publishers, Dordrecht, The Netherlands, pp 253–264

Llamas R, Martinez-Santos P (2005) Intensive groundwater use: silent revolution and potential source of social conflicts. J Water Resour Plann Manage 131:337–341

Lopez-Gunn E (2009) Agua para todos: a new regionalist hydraulic paradigm in Spain. Water Altern 2(3):370–394

Lynch A, Tryhorn L, Abramson R (2008) Working at the boundary facilitating interdisciplinarity in climate change adaptation research. Am Meteorol Soc, 169–179. doi:10.1175/BAMS-89-2-169

Madzwamuse M (2010) Climate governance in Africa: adaptation strategies and institutions. Heinrich-Böll-Foundation, Berlin

Magnan A et al (2009) The future of the Mediterranean: from impacts of climate change to adaptation issues. IDDRI, 1–41

MAPMADA. Ministère de l'Agriculture et de la Pêche Maritime et Agence pour le Développement Agricole (2008) Le Plan Maroc Vert. s.n., s.l.

Mariotti A et al (2008) Mediterranean water cycle changes: transition to drier 21st century conditions in observations and CMIP3 simulations. Environ Res Lett, 3, s.l.

Martin R, Linstädter A, Frank K, Müller B (2016) Livelihood security in face of drought—assessing the vulnerability of pastoral households. Environ Model Softw, 414–423

MEMEE, Ministère de l'Energie, des Mines, de l'Eau et de l'Environnement (2010) Rapport National sur l'État de l'Environnement. s.n., s.l.

Méndez J (2008) La adaptación al Cambio Climático en Andalucía. s.n., Madrid

Mestre-Barceló A (1995) Five-year drought continues in Spain. Drought Network News (1994–2001). University of Nebraska

MINETUR, Ministerio de Industria, Energía y Turismo (2013) Plan Nacional e Integral de Turismo (PNIT) 2012–2015. s.n., s.l.

Molina F, Villa A (2008) La Reserva de Biosfera Intercontinental de Mediterráneo Andalucía (España)—Marruecos como instrumento de cooperación. Ecosistemas 17(2):17–27

Moore H, Fox H, Harrouni M, El Alami A (1998) Environmental challenges in the Rif Mountains, Northern Morocco. Environ Conserv 25:354–365

Moreira J, Ribalaygua J (2007) Escenarios Actuales y Futuros del Clima para Andalucía. Consejería de Medio Ambiente, Junta de Andalucía. s.n., s.l.

Moriasi D et al (2007) Model evaluation guidelines for systematic quantification of accuracy in watershed simulations. Trans ASABE 50(3):885–900

Moriondo M, Bindi M (2007) Impact of climate change on the phenology of typical Mediterranean crops. Ital J Agrometeorol 3:5–12

Moriondo M et al (2006) Potential impact of climate change on fire risk in the Mediterranean area. Clim Resour 31:85–95

Murray C, Rosenfeld L, Lim S, Andrews K, Foreman K, Haring D et al (2012) Global Malaria mortality between 1980 and 2010: a systematic analysis. Lancet 379:413–431

Navarra A, Tubiana L (2013) Regional assessment of climate change in the Mediterranean. Springer, Dordrecht

Nicholls R, Hoozemans F (1996) The Mediterranean: vulnerability to coastal implications of climate change. Ocean Coast Manage 31(2–3):105–132

Nunes J, Seixas J, Pacheco N (2008) Vulnerability of water resources, vegetation productivity and soil erosion to climate change in Mediterranean watersheds. Hydrol Process 16:3115–3134

OECD (2007) African economic outlook. s.n., Paris

Otero M, Garrabou J, Vargas M (2013) Mediterranean Marine Protected Areas and climate change: a guide to regional monitoring and adaptation opportunities. IUCN, Malaga

Ouassou A et al (2005) Analysis of the organizations and institutions, in drought preparedness and mitigation in the Mediterranean. Centre International de Hautes Études Agronomiques Méditerranéennes (CIHEAM), Morocco

Oxford Business Group (2009) Morocco 2009 (Report). s.n., s.l.

Parish R, Funnell D (1999) Climate change in mountain regions: some possible consequences in the Moroccan High Atlas. Glob Environ Change 9:45–58

Parry M et al (2007) Climate change 2007: impacts, adaptation, and vulnerability. Contribution of WGII to the fourth assessment report of the IPCC. Cambridge University Press, s.l.

Peñuelas J, Lloret F, Montoya R (2001) Severe drought effects on Mediterranean woody flora in Spain. For Sci 47(2)

Perez, A., Gil, S., Lopez, F. & Barriendos, M., 2016. Increasing severity of damage caused by floods in the Spanish Mediterranean coast (1960–2014), climate change or vulnerability? Geophysical Research Abstracts, s.l., p 4191

Pinkerton K, Rom W (2014) Global climate change and public health. Respiratory Medicine Series. Springer Science + Business Media, New York

Pla E, Pascual D, Ordóñez J (2012) Adaptacions al Canvi Climàtic en l'Ús de l'Aigua. Memòria divulgativa del projecte ACCUA. Obra Social CatalunyaCaixa

Plan Bleu (2008) Climate change and energy in the Mediterranean. European Investment Bank, Sophia Antipolis

Pryor S, Howe J, Kunkel K (2009) How spatially coherent and statistically robust are temporal changes in extreme precipitation in the contiguous USA? Int J Climatol 45:31–45

Refass M (1992) Historical migration patterns in the eastern Rif mountains. Mt Res Dev 12: 383–388

RME, Regional Ministry of the Environment of Spain (2008) Environment in Andalusia: report summary. s.n., Spain

RMSI (2012) Morocco natural hazards probabilistic risk analysis and national strategy development drought hazard report. Department of Economic and General Affairs, Kingdom of Morocco, s.l.

Roberts C (2002) Drought management in the Río Guadalhorce region of Andalucía, southern Spain. Land Degrad Dev, 151–163

Roudies N (2013) Presentation on ecotourism, poverty reduction and environmental protection. 29 Oct 2013. Secretary General, Moroccan Ministry of Tourism Expert Group. s.n., s.l.

Ruiz Sinoga J, Romero Díaz A, Ferre Bueno E, Martínez Murillo J (2010) The role of soil surface conditions in regulating runoff and erosion processes on a metamorphic hillslope (Southern Spain): soil surface conditions, runoff and erosion in Southern Spain. Catena 80:131–139

Sarris D, Christodoulakis D, Körner C (2007) Recent decline in precipitation and tree growth in the eastern Mediterranean. Glob Change Biol 13:1187–1200

Scheffran J, Battaglini A (2011) Climate and conflicts: the security risks of global warming. Reg Environ Change 11(1):27–39

Schnepf D, Lutter F (2012) Integrated participatory water resources management and water governance. In: Integrated water resources management in the mediterranean Region. Dialogue towards new strategy. s.n., s.l., pp 313–329

Schröter D, Cramer W, Leemans R et al (2005) Ecosystem service supply and human vulnerability to global change in Europe. Science 310(5732):1333–1337

Senatore A, Mendicino G, Smiatek G, Kunstmann H (2011) Regional climate change projections and hydrological impact analysis for a Mediterranean basin in Southern Italy. J Hydrol 399:70–92

Serra-Diaz J, Ninyerola M, Lloret F (2012) Coexistence of *Abies alba* (Mill.)—*Fagus sylvatica* (L.) and climate change impact in the Iberian Peninsula: a climatic-niche perspective approach. Funct Ecol Plants 207(1):10–18

Simonet S (2011) Adapting to climate change in the water sector in the Mediterranean: situation and prospects. Blue Plan Papers, s.l.

Skees J et al (2001) Developing rainfall-based index insurance in Morocco. In: Policy Research Working paper 2577. The World Bank, Washington DC

Snoussi M, Ouchani T, Niazi S (2008) Vulnerability assessment of the impact of sea-level rise and flooding on the Moroccan coast: the case of the Mediterranean eastern zone. Estuar Coast Shelf Sci 77:206–213

Spanish National Report (2016) 28th Session of the International Coordinating Council of the Man and Biodiversity. Lima, Peru

Stern N (2006) Review on the economics of climate change. s.n: s.l.

Swart RJ, Biesbroek GR, Binnerup S, Carter TR, Henrichs T, Loquen S et al (2009) Europe adapts to climate change: comparing national adaptation strategies (No. 01/2009). Finnish Environment Institute (SYKE), Helsinki

Tàbara J, Ilhan A (2008) Culture as a trigger for sustainability transition in the water domain. Reg Environ Change 8(2):59–71

Tellería J, Fernández-López J, Fandos G (2016) Effect of climate change on mediterranean winter ranges of two migratory passerines. PLoS ONE 11(1)

The Amsterdam Declaration on Global Change (2001) Challenges of a changing earth: global change open science conference. s.n., s.l.

Thuiller W et al (2005) Climate change threats to plant diversity in Europe. Proc Natl Acad Sci U S A 102:8245–8250

Tielbörger K et al (2014) Middle-Eastern plant communities tolerate 9 years of drought in a multi-site climate manipulation experiment. Nat Commun 5(5102)

Touchan R et al (2010) Spatiotemporal drought variability in northwestern Africa over the last nine centuries. Clim Dyn, pp 237–252

Troccoli A (2008) Seasonal climate: forecasting and managing risk. Springer Science & Business Media, s.l.

Turco M, Llasat M (2011) Trends in indices of daily precipitation extremes in Catalonia (NE Spain), 1951–2003. Nat Hazards Earth Syst Sci 11:3213–3226

UNDESA—United Nations, Department of Economic and Social Affairs, Population Division (2013) International Migration Report 2013. s.n., s.l.

UNECE (2014) Morocco: environmental performance reviews. UNECE/UNECA, New York and Geneva

UNEP-MAP RAC/SPA (2009) Sub-regional report on vulnerability and impacts of climate change on marine and coastal biological diversity in the Mediterranean, Arab Countries, Tunis: By Ben Haj S, Cebrian D, Limam A, Grimes S, Halim Y, Bitar G, Bazairi H, Ibrahim A, Romdhane MS (ed) RAC/SPA

UNFCCC (2007) Climate change: impacts, vulnerabilities and adaptation in developing countries. s.n., Bonn

UNFCCC Secretariat (2008) Compendium on methods and tools to evaluate impacts of, and vulnerability and adaptation to, climate change. Feb 2008. s.n., s.l.

UNU-EHS, United Nations Country Team (2007) Millennium development goals—Morocco. Haut-Commisaria du Plan, s.l.

UNWTO (2015) Tourism highlights. 2015 Edition, s.l.

Vicente-Serrano SM, Gouveia C, Camarero JJ et al (2013) The response of vegetation to drought time-scales across global land biomes. Proc Natl Acad Sci U S A 110(1):52–57

Vogt J, Somma F (2000) Drought and drought mitigation in Europe. Kluwer Academic Publishers, Dordrecht

WHO/UNICEF (2015) Estimates on the use of water sources and salination facilities in Morocco. Joint Monitoring Programme for Water Supply and Sanitation, s.l.

Wilhite D, Glantz M (1985) Understanding the drought phenomenon: the role of definitions. Water Int 10(3):111–120

Wilhite D, Vanyarkho O (2000) Drought: pervasive impacts of a creeping phenomenon (chapter 18). In: Drought: a global assessment. Natural Hazards and Disasters Series. s.n., Routledge Publishers

World Bank (2010) World Development Report 2010: development and climate change. Washington DC, World Bank

World Bank (2011) Integrating climate change into the Plan Maroc Vert (PMV) project. World Bank, Washington DC

World Bank (2012) Turn down the heat: why a 4 °C warmer world must be avoided. World Bank, Washington DC

World Bank (2014) Natural disasters in the Middle East and North Africa: a regional overview. World Bank: s.n., Washington DC

World Bank Group (2014) Turn down the heat: confronting the new climate normal. World Bank, Washington DC

Zamora J, Verdú J, Galante E (2007) Species richness in Mediterranean agroecosystems: spatial and temporal analysis for biodiversity conservation. Biol Conserv 134:113–121

Zghibi A, Zouhri L, Tarhouni J, Kouzana L (2013) Groundwater mineralisation processes in Mediterranean semi-arid systems (Cap-Bon, North east of Tunisia): hydrogeological and geochemical approaches. Hydrol Process 27(22):3227–3239

Zubrycki K, Crawford A, Hov H, Parry J (2011) Review of current and planned adaptation action: North Africa. International Institute for Sustainable Development, Geneva